Life Science
Grade Three
Table of Conte

Life Science
Grade Three
Introduction

We wake up in a new world every day. Our lives are caught in a whirlwind of change in which new wonders are constantly being discovered. Technology is carrying us headlong into the 21st century. How will our children keep pace? We must provide them with the tools necessary to go forth into the future. Those tools can be found in a sound science education. One guidepost to a good foundation in science is the National Science Education Standards. This book adheres to these standards.

Young children are naturally curious about science and life. They see the world around them and ask questions that naturally lead into the lessons that they will be taught in science. Science is exciting to children because it answers their questions about themselves and the world around them—their immediate world and their larger environment. A basic understanding of science boosts students' understanding of the world around them.

As children learn more about themselves and their world, they should be encouraged to notice the other living things that inhabit their world. They should be aware of the life cycles of all living things. They should become aware of the interdependence of organisms—from plants, to animals, to humans—and how organisms affect and are affected by their environments. Children should also learn how they can control their own environments to promote their health. Through good personal hygiene, exercise, and making good decisions while interacting with nature, children will learn how to take better care of themselves.

Organization

Life Science serves as a handy companion to the regular science curriculum. It is broken into four units: Living and Non-Living Things, Plants, Animals, and Health. Each unit contains concise background information on the unit's topics, as well as exercises and activities to reinforce students' knowledge and understanding of basic principles of science and the world around them.

- **Living and Non-Living Things.** Students are introduced to the differences between living and non-living things. They learn the basic makeup of animal and plant cells and how cells grow and divide. Students are shown the relationship between growing cells and growing organisms. Students become familiar with single-celled organisms.

- **Plants.** Students learn about plants and what they need to survive. They experiment with deprivation of these needs and see the results. Students learn how different plants grow and reproduce.

- **Animals.** Students work with classification of different animals. They study the needs of animals and animal behaviors and adaptations that ensure that these needs are met. Students learn about the birth and care of young animals. Students become familiar with the many different types of environments on the Earth and the animals that are suited to living in each.

- **Health.** Students learn about nutrition, exercise, and hazards in the home and their environments. They explore ways in which they can make good choices and find that they have control over their own health. They learn healthy habits that they can develop and keep for life.

This book contains three types of pages:

- Concise background information is provided for each unit. These pages are intended for the teacher's use or for helpers to read to the class.
- Exercises are included for use as tests or practice for the students. These pages are meant to be reproduced.
- Activity pages list the materials and steps necessary for students to complete a project. Questions for students to answer are also included on these pages as a type of performance assessment. As much as possible, these activities include most of the multiple intelligences so students can use their strengths to achieve a well-balanced learning style. These pages are also meant for reproduction for use by students.

Use

Life Science is designed for independent use by students who have been introduced to the skills and concepts described. This book is meant to supplement the regular science curriculum; it is not meant to replace it. Copies of the activities can be given to individuals, pairs of students, or small groups for completion. They may also be used as a center activity. If students are familiar with the content, the worksheets may also be used as homework.

To begin, determine the implementation that fits your students' needs and your classroom structure. The following plan suggests a format for this implementation.

1. Explain the **purpose** of the worksheets to your students. Let them know that these activities will be fun as well as helpful.

2. Review the **mechanics** of how you want the students to work with the activities. Do you want them to work in groups? Are the activities for homework?

3. Decide how you would like to use the **assessments.** They can be given before and after a unit to determine progress, or only after a unit to assess how well the concepts have been learned.

4. Determine whether you will send the tests home or keep them in students' **portfolios.**

5. Introduce students to the **process** and the purpose of the activities. Go over the directions. Work with children when they have difficulty. Work only a few pages at a time to avoid pressure.

6. Do a **practice** activity together.

The Scientific Method

Students can be more productive if they have a simple procedure to use in their science work. The scientific method is such a procedure. It is detailed here, and a reproducible page for students is included on page 7.

1. **PROBLEM:** Identify a problem or question to investigate.
2. **HYPOTHESIS:** Tell what you think will be the result of your investigation or activity.
3. **EXPERIMENTATION:** Perform the investigation or activity.
4. **OBSERVATION:** Make observations, and take notes about what you observe.
5. **CONCLUSION:** Draw conclusions from what you have observed.
6. **COMPARISON:** Does your conclusion agree with your hypothesis? If so, you have shown that your hypothesis was correct. If not, you need to change your hypothesis.
7. **PRESENTATION:** Prepare a presentation or report to share your findings.
8. **RESOURCES:** Include a list of resources used. Students need to give credit to people or books they used to help them with their work.

Hands-On Experience

An understanding of science is best promoted by hands-on experience. *Life Science* provides a wide variety of activities for students. But students also need real-life exposure to their world. Playgrounds, parks, and vacant lots are handy study sites to observe many organisms. Repeated visits to the same site can help to show students that the organisms are constantly changing.

It is essential that students be given sufficient concrete examples of scientific concepts. Appropriate manipulatives can be bought or made from common everyday objects. Most of the activity pages can be completed with materials easily accessible to the students. Manipulatives that can be used to reinforce scientific skills are recommended on several of the activity pages.

Science Fair

Knowledge without application is wasted effort. Students should be encouraged to participate in their school science fair. To help facilitate this, each unit in *Life Science* ends with a page of science fair ideas and projects. Also, on page 8 is a chart that will help students to organize their science fair work.

To help students develop a viable project, you might consider these guidelines:

- Decide whether to do individual or group projects.

- Help students choose a topic that interests them and that is manageable. Make sure a project is appropriate for a student's grade level and ability. Otherwise, that student might become frustrated. This does not mean that you should discourage a student's scientific curiosity. However, some projects are just not appropriate. Be sure, too, that you are familiar with the school's science fair guidelines. Some schools, for example, do not allow glass or any electrical or flammable projects. An exhibit also is usually restricted to three or four feet of table space.

- Encourage students to develop questions and to talk about their questions in class.

- Help students to decide on one question or problem.

- Help students to design a logical process for developing the project. Stress that the acquisition of materials is an important part of the project. Some projects also require strict schedules, so students must be willing and able to carry through with the process.

- Remind students that the scientific method will help them to organize their thoughts and activities. Students should keep track of their resources used, whether they are people or print materials. Encourage students to use the Internet to do research on their project.

Additional Notes

- **Parent Communication:** Send the Letter to Parents home with students so that parents will know what to expect and how they can best help their child.

- **Bulletin Board:** Display completed work to show student progress.

- **Portfolios:** You may want your students to maintain a portfolio of their completed exercises and activities, or of newspaper articles about current events in science. This portfolio can help you in performance assessment.

- **Assessments:** There are Assessments for each unit at the beginning of the book. You can use the tests as diagnostic tools by administering them before children begin the activities. After children have completed each unit, let them retake the unit test to see the progress they have made.

- **Center Activities:** Use the worksheets as a center activity to give students the opportunity to work cooperatively.

- **Have fun.** Working with these activities can be fun as well as meaningful for you and your students.

CURRICULUM CORRELATION

Curriculum Area	Page Numbers
Social Studies	87, 89
Language Arts	15, 16, 20, 21, 22, 25, 29, 31, 33, 34, 35, 39, 43, 44, 45, 47, 48, 53, 56, 66, 68, 74, 75, 76, 77, 79, 80, 84, 86, 88, 90, 92, 93, 97, 99, 100, 102, 110, 125, 126, 130, 132,135, 136, 140
Math	23, 32, 41, 42, 45, 91, 92, 122, 123, 124, 126, 128, 129, 130
Physical Education/Health	109, 110, 111, 112, 113, 114, 115, 116, 117, 118, 119, 120, 121, 122, 123, 124, 125, 126, 127, 128, 129, 130, 131, 132, 133, 134, 135, 136, 137, 138, 139, 140, 141
Art	18, 19, 22, 24, 26, 30, 46, 47, 50, 51, 52, 57, 73, 80, 82, 85, 89, 96, 98, 131, 133

FOSS CORRELATION

The Full Option Science System™ (FOSS) was developed at the University of California at Berkeley. It is a coordinated science curriculum organized into four categories: Life Science; Physical Science; Earth Science; and Scientific Reasoning and Technology. Under each category are various modules that span two grade levels. The modules for this grade level are highlighted in the chart below.

Module	Page Numbers
Human Body	13-15, 22, 59-61, 70, 106-109, 110, 111-112, 113-114, 117-118, 119, 120, 122, 123-124, 126-126, 127, 128, 129-130, 131, 132, 133, 140, 141
Structures of Life	13-15, 16, 17, 18, 19, 20, 21, 22, 23, 24, 25, 26, 27, 28, 29, 30, 31, 32, 33, 34, 35, 37-38, 39, 40, 41-42, 43-44, 45-46, 47, 48, 49, 50, 51, 52, 53-54, 55, 56, 57, 59-61, 62, 63-64, 65-66, 67-68, 69, 70, 71-72, 73-74, 75-76, 77, 78, 79-80, 81-82, 83-84, 85-86, 87, 88, 89-90, 91-92, 93-94, 95-96, 97, 98, 99-100, 101-102, 103-104, 134, 135-136, 137-138, 139

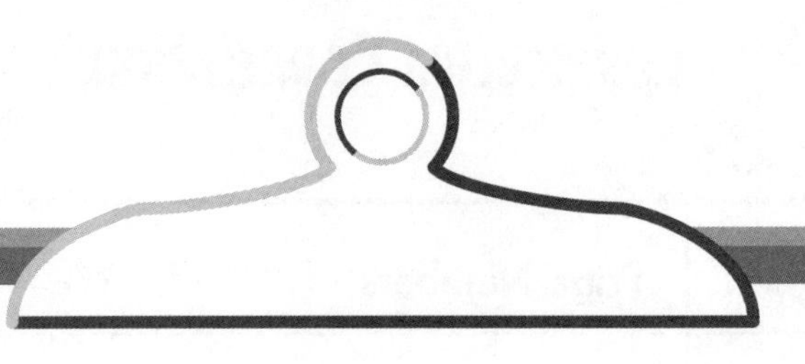

Dear Parent,

During this school year, our class will be using an activity book to reinforce the science skills that we are learning. By working together, we can be sure that your child not only masters these science skills but also becomes confident in his or her abilities.

From time to time, I may send home activity sheets. To help your child, please consider the following suggestions:

- Provide a quiet place to work.
- Go over the directions together.
- Help your child to obtain any materials that might be needed.
- Encourage your child to do his or her best.
- Check the activity when it is complete.
- Discuss the basic science ideas associated with the activity.

Help your child to maintain a positive attitude about the activities. Let your child know that each lesson provides an opportunity to have fun and to learn more about the world around us. Above all, enjoy this time you spend with your child. As your child's science skills develop, he or she will appreciate your support.

Thank you for your help.

Cordially,

Name ______________________________ Date ______________

THE SCIENTIFIC METHOD

Did you know you think and act like a scientist? You can prove it by following these steps when you have a problem. These steps are called the scientific method.

1. **PROBLEM:** Identify a problem or question to investigate.

   ____

   ____

2. **HYPOTHESIS:** Tell what you think will be the result of your investigation or activity.

   ____

   ____

3. **EXPERIMENTATION:** Perform the investigation or activity.

   ____

4. **OBSERVATION:** Make observations, and take notes about what you observe.

   ____

   ____

   ____

5. **CONCLUSION:** Draw conclusions from what you have observed.

   ____

   ____

   ____

6. **COMPARISON:** Does your conclusion agree with your hypothesis? If so, you have shown that your hypothesis was correct. If not, you need to change your hypothesis.

   ____

7. **PRESENTATION:** Prepare a presentation or report to share your findings.
8. **RESOURCES:** Include a list of resources used. You need to give credit to people or books you used to help you with your work.

Name ______________________________ Date ______________

The Science Fair

The science fair at your school is a good place to show your science skills and knowledge. Science fair projects can be several different types. You can do a demonstration, make a model, present a collection, or perform an experiment. You need to think about your project carefully so that it will show your best work. Use the scientific method to help you to organize your project. Here are some other things to consider:

Project Title ______________________________

Working Plan	Date Due	Date Completed	Teacher Initials
1. Select topic			
2. Explore resources			
3. Start notebook			
4. Form hypothesis			
5. Find materials			
6. Investigate			
7. Prepare results			
8. Prepare summary			
9. Plan your display			
10. Construct your display			
11. Complete notebook			
12. Prepare for judging			

Write a brief paragraph describing the hypothesis, materials, and procedures you will include in your exhibit. Be sure to plan your project carefully. Get all the materials and resources you need beforehand. Also, a good presentation should have plenty of visual aids, so use pictures, graphs, charts, and other things to make your project easier to understand.

Be sure to follow all the rules for your school science fair. Also, be prepared for the judging part. The judges will look for a neat, creative, well-organized display. They will want to see a clear and thorough presentation of your data and resources. Finally, they will want to see that you understand your project and can tell them about it clearly and thoroughly. Good luck!

Name ______________________________ Date ________________

WHAT DO YOU THINK?

Fill in the blanks with the correct words from the list.

Living things are not like **(1)** ________________ things. Living things need **(2)** ________________ in order to grow. Animals can **(3)** ________________ from place to place. Plants cannot, but they can turn their

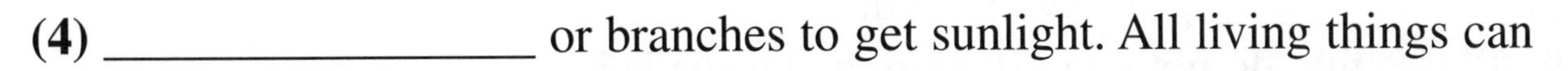

(4) ________________ or branches to get sunlight. All living things can **(5)** ________________ to make other living things like themselves.

All living things are made of **(6)** ________________. Plant cells are different from animal cells. Plant cells have a cell **(7)** ________________. Animal cells do not. Both animal cells and plant cells are surrounded by a cell **(8)** ________________. This holds the cell together. The control center for a cell is its **(9)** ________________. It directs the growth and **(10)** ________________ of the cell. The nucleus floats in a jelly-like liquid cell **(11)** ________________.

WORD LIST		
non-living	cells	walls
leaves	nucleus	food
reproduce	move	cytoplasm
reproduction	membrane	

Name ______________________________ Date ______________

PLANT CROSSWORD PUZZLE

Use the words in the box to answer the clues. Then fill in the puzzle.

fungi	ovule	pistil	pollinate	cones
spores	pollen	stamen	chlorophyll	

ACROSS

3. The green material in plants is ________.
6. The grains of yellow powder on the stamen are ________.
7. Plants without chlorophyll are ________.
8. The ________ produces pollen.
9. Plants that do not have flowers make seeds in ________.

DOWN

1. Pollen must fall in the ________ of the flower.
2. The ________ of a flower grows in a seed.
4. To ________ a flower, pollen from the stamen must land on the pistil.
5. New mushrooms grow from ________.

1 2 3 4 5 6 7 8 9

Name ______________________________ Date ______________

ANIMAL SENSE

Choose the correct word to complete each sentence and write it in the blank.

1. Features that living things have are called ____________.
   faces traits groups

2. Animals are classified into two large groups by dividing those who have ________________ from those who do not.
   wings fur backbones

3. ________________ live part of their lives on water and part on land.
   Amphibians Mammals Birds

4. Giving birth to live young and having fur are traits of _____________.
   reptiles amphibians mammals

5. An animal's ________________ are suited to the kind of food it must eat.
   legs claws teeth

6. Birds have many different types of ________________ to help them to get their food.
   claws beaks feathers

7. Animals must ________________ to their environment to survive.
   adapt run swim

8. Behaviors that do not have to be learned are called _____________ .
   camouflage defense instincts

9. Some animals use ________________ to blend into their surroundings.
   instincts odors camouflage

10. Flying south in the winter is called ________________.
    hibernation migration protection

Name ________________________ Date ______________

Good Health

Choose a word from the box that best fits each clue. Write the word in the shapes.

WORD LIST				
grains	dentist	mineral	teeth	sneeze
germs	pollen	smog	sunburn	ivy

1. Cereals, bread, and pasta are these. ○□□□□□
2. Wash these off to stay healthy. □○□□□
3. This person will help your teeth. ○□□□□□□
4. This floats in the air and causes illness. □○□□□□
5. This plant will give you a rash. **Poison** □□○
6. Wear sunblock to avoid this. □□○□□□□
7. Calcium is one that builds strong bones. □○□□□□□
8. This is smoke trapped near the Earth. □□□○
9. Cover your mouth or nose when it happens. □□○□□□
10. Take care of these for a healthy smile. □□□□○

Challenge: Arrange the letters in the circles to spell two words that will help you stay healthy. Two letters are already in place to help you start. Some letters may be used more than once.

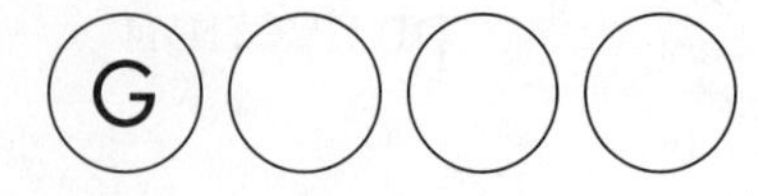

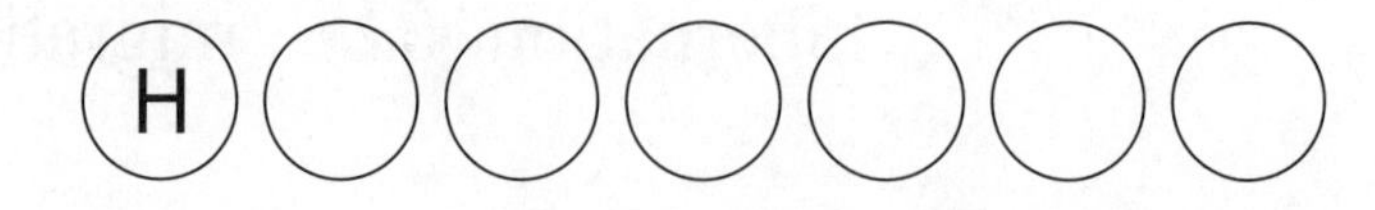

Unit I: Living and Non-Living Things
Background Information

Living/Non-Living

All living things carry on activities that non-living things do not. These life processes define a living thing. All living things grow, or increase in size and the amount of matter they contain. All living things can reproduce, or make more of the same kind of organism. Living things consume energy, change it, and excrete, or give off, waste. Living things react to stimuli and to changes in the environment.

Non-living things may carry on some of these activities, but because they do not carry on all of these activities, they are not living. Students may be confused about what is living and what is not. Water seems to move, change, and appear alive. A flame will flicker and grow. Even scientists disagree about certain things, such as viruses. Distinguishing between living and non-living things can be difficult, but students can follow the guidelines above to form a strong grasp on the concept.

Cells

Living things are made up of cells. A cell is the smallest living unit. It has all of the properties of a living thing. Most cells contain a nucleus, cytoplasm, and a cell membrane. Plant cells have, in addition, a cell wall, which is outside of the membrane. The cell wall makes the plant cell stiffer than the animal cell. A tree trunk is hard because of the cell walls in the plant cells. The crunch of a carrot or celery is caused by the cell walls breaking as we bite into the food.

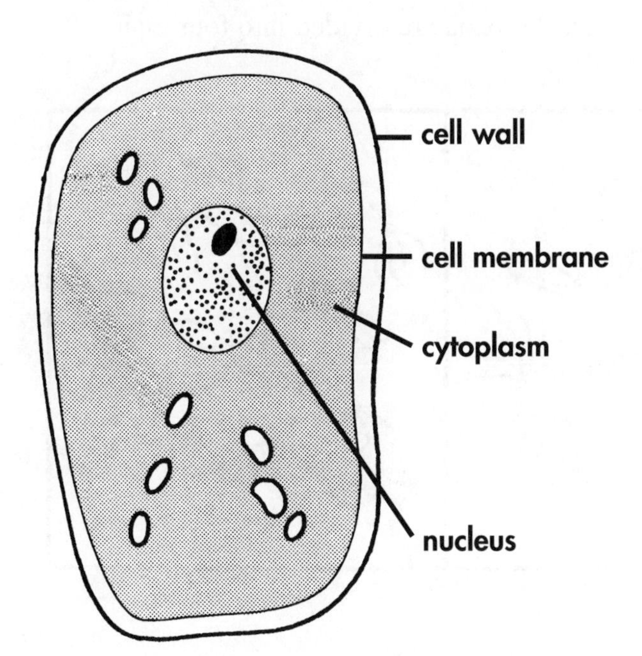

The nucleus of a cell contains the genetic information, or that information that will determine the makeup of the organism. All life processes occur within the cell. Living cells produce new living cells. The cytoplasm floats inside the membrane, holding small pieces of living matter. Within the cytoplasm are sometimes small pockets called vacuoles. These small spaces store food. The cell membrane and cell wall hold the cell together and give it shape. Food and oxygen are allowed to pass through the cell membrane, and wastes are excreted through the membrane as well. Although all cells have the same basic elements, not all cells look exactly alike; they vary in size and shape. For example, humans have several types of cells that perform different functions in the body. Blood cells, which move around constantly, are round and well suited to movement. Muscle cells are long and narrow like the elastic muscles they form.

Students may observe a cell by looking through a microscope. A transparent piece of onion skin on a slide with a drop of iodine will show students a plant cell. A scraping from the inside of the cheek (with a toothpick) mixed in with a drop of iodine on a slide will give students the opportunity to observe an animal cell.

Growth

The cells of most organisms are about the same size. What makes an elephant larger than a mouse is the number of cells contained in the elephant, not the size of each cell. Most cells reach a certain size and stop growing. When they reproduce, the new cells are smaller, but then they grow, too.

Before a cell divides, it enters a stage called interphase during which it digests food, uses it for energy, and excretes waste. During this time the cell grows in preparation for division. Different cells divide at different rates, from 15 to 30 minutes to almost two days.

The division of cells is called *mitosis.* There are four stages of cell division.

- The first, prophase, is when the chromosomes in the nucleus become shorter and fatter, then duplicate themselves. The membrane of the nucleus begins to break down.
- During the second phase, metaphase, the chromosomes line up across the nucleus of the cell. Each chromosome begins to pull apart, separating the duplicated information from the original information.
- At anaphase, the third phase, the chromosomes pull toward opposite ends of the cell.
- During the last phase, telophase, cytoplasm is divided, and the nucleus reorganizes into two cells.

The growth of a living thing is caused by the growth and division of its cells.

Single-Celled Organisms

Single-celled organisms, or microorganisms, such as bacteria, viruses, and various single-celled animals and plants, consist of only one cell.

Bacteria are among the oldest organisms known. They live in the soil where they help to break down dead materials and aid in decay. Many bacteria also live in animals and humans, in the gut, and on the skin, where they cause no harm. Bacteria are classified by shape—round, spiral, or rod-like. Only a few bacteria cause diseases. Bacteria have a simple structure, and are classified as neither plants nor animals.

Protozoa are a group of tiny animals found wherever there is enough moisture to support life. Protozoa are divided into four main

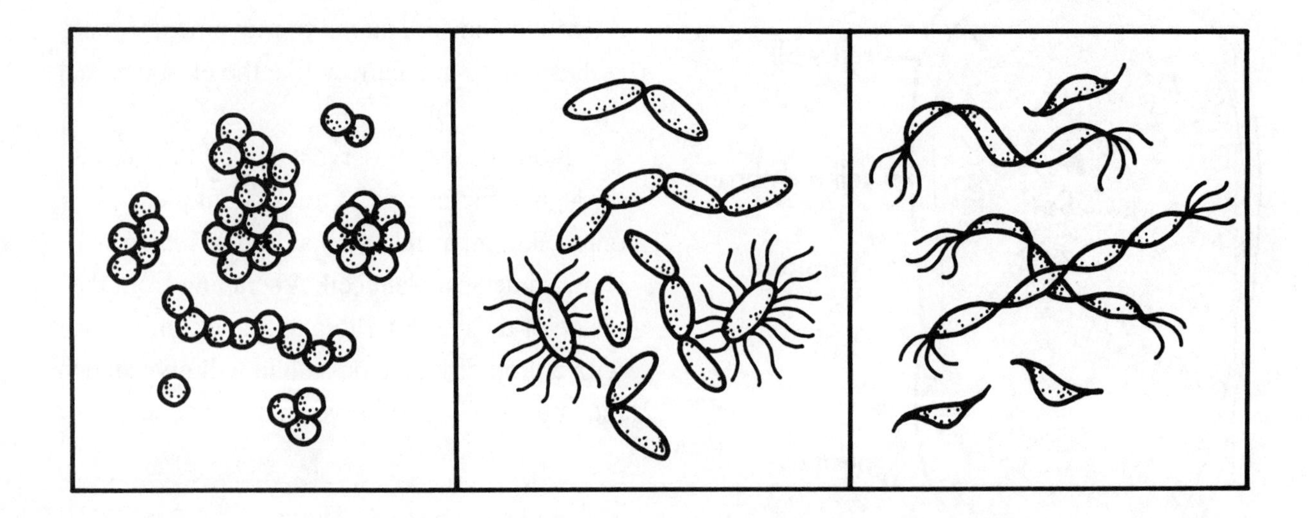

groups: those with flagella, spidery "tails" which are used for movement and for eating; those with cilia, tiny moving hairs that surround the cell; the amoeba, which has no real shape but constantly changes shape as it moves; and those that cannot move independently.

Most plants make their own food. Algae make their food from non-living things. Plant cells have chloroplasts, which trap energy from the sun. Water and carbon dioxide enter the cell through the cell wall. The cell turns the water and gas into food and oxygen. The cell uses the food and passes off the oxygen to be used by other living things.

A fungus is a plant-like organism that cannot make its own food because it has no chloroplasts. A fungus lives on things that are, or were, alive. Yeast and mold are fungi that grow and reproduce very quickly. Some molds ruin food, others actually give good flavor, such as those used in making blue cheese. Yeast is used in bread making; the gases given off by the yeast cause the holes in the bread.

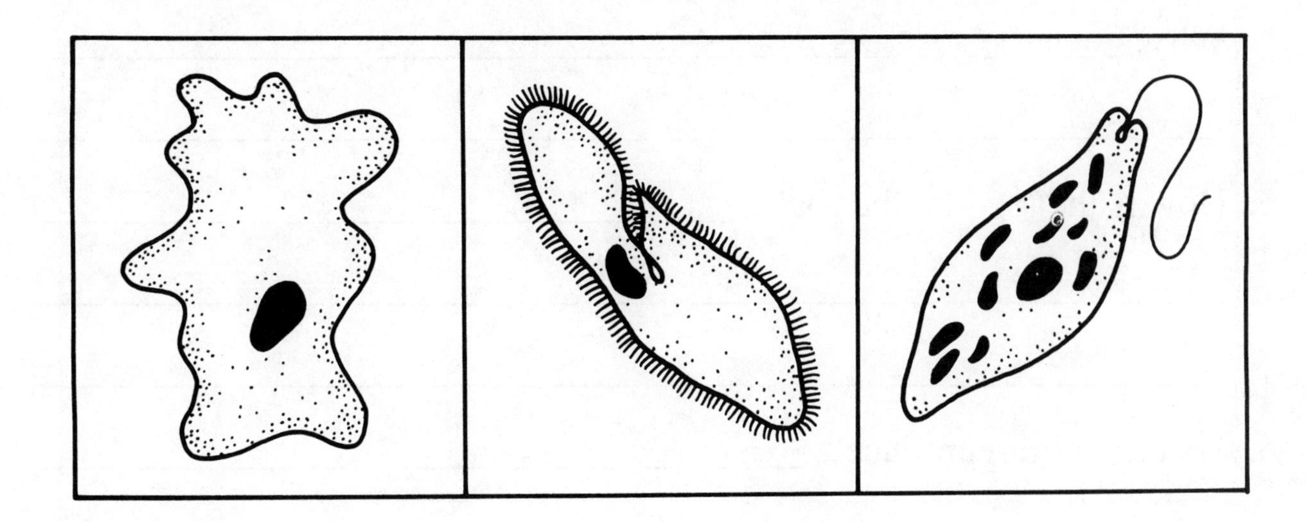

Name ______________________ Date ____________

Is It Alive?

In the picture are a tree, a log, a bird, and a rock. The tree is growing. The log was cut from a tree.

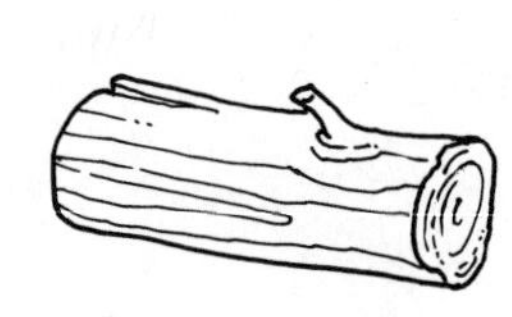

Answer these questions.

1. Are the tree and the log both alive? ______________________

Why do you say so? ______________________

____

2. What does a tree need to live? ______________________

3. What does the bird need to live? ______________________

4. Does the rock need these things, too? ______________________

Why? ______________________

____

5. How is the log like the rock? ______________________

____

____

6. Will the bird reproduce? ______________________

____

Name ______________________________ Date ______________

LIVING AND NON-LIVING THINGS

A. Put a circle around the pictures that show living things.

   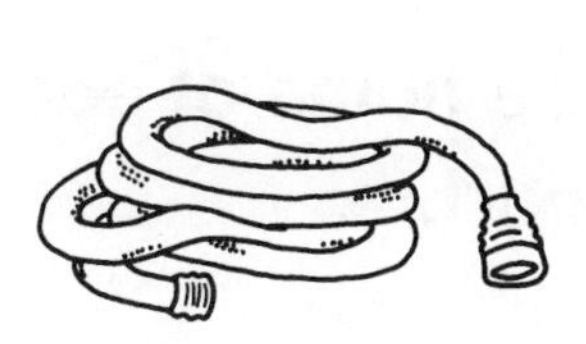 

B. Read the words below. Then put a square around those that help us tell a living thing from a non-living thing.

moves by itself	has color	reproduces	has weight
needs food	grows	crumbles	is warm

Name ______________________________ Date ______________

WHAT ARE LIVING THINGS MADE OF?

Living things are made of many tiny units called *cells.* Plant cells and animal cells are not exactly alike. Plant cells have a cell wall. The cell wall surrounds the cell membrane. The center of a cell is its nucleus. The liquid in a cell is called cytoplasm.

A. Color the diagram of a plant cell.

1. Color the cell wall green.
2. Color the cell membrane red.
3. Color the nucleus brown.
4. Color the cytoplasm blue.

B. Write the name of each part of the cell on the line that points to the part.

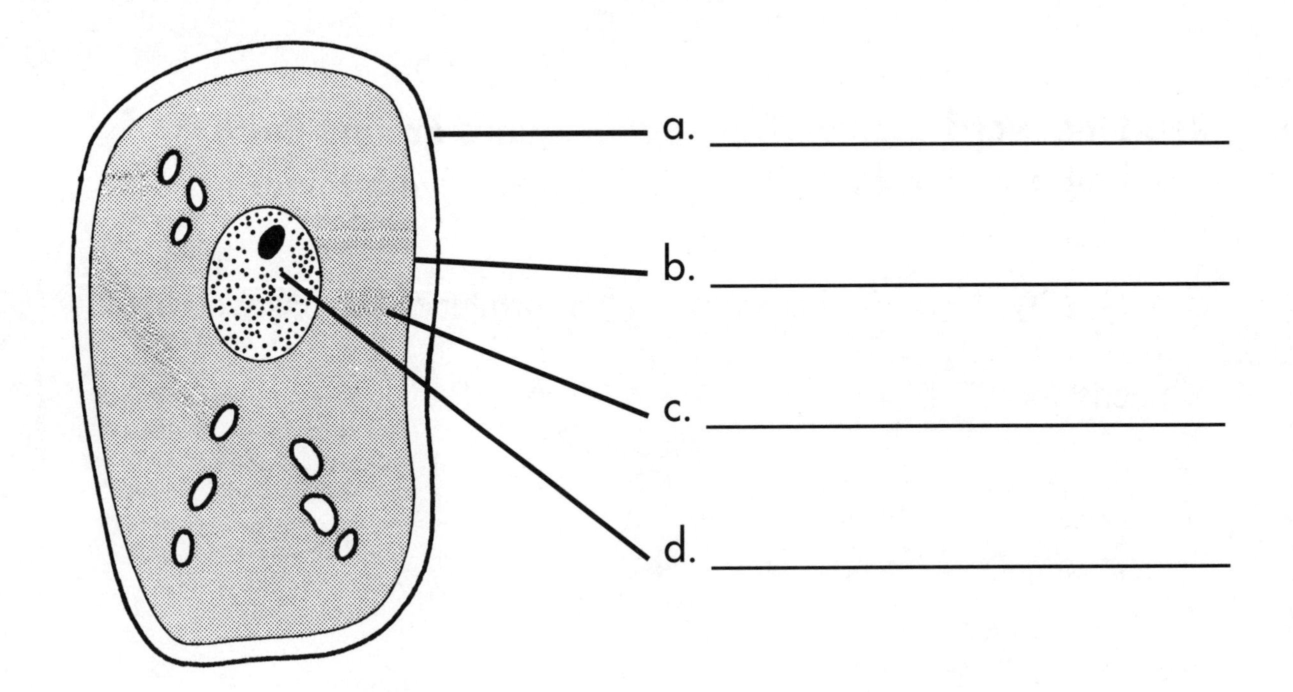

Name ______________________________ Date ______________

DRAWING A CELL

A. The parts of a cell are listed below. Fill in the blanks with the correct words from the list.

cytoplasm	**nucleus**
cell wall	**cell membrane**

1. You can tell a plant cell from an animal cell because it has a ____________________.

2. The thin, skinlike covering that surrounds the cell is the ______________________________.

3. The living liquid inside the cell is the ____________________.

4. The control center of a living cell is the ____________________.

B. Draw a cell in the space below. Using the above sentences, number each part.

Name ______________________ Date ______________

A Home Egg-speriment

At school, you are learning that living things are made of cells. In this at-home activity, you will look at the parts of a large cell. THIS ACTIVITY MUST BE SUPERVISED BY AN ADULT.

A. **Gather these materials: 1 chicken egg and 1 bowl.**
B. **Crack the egg and place it in the bowl. Do not break the yellow center.**
C. **Using the picture below, find the parts of the cell.**

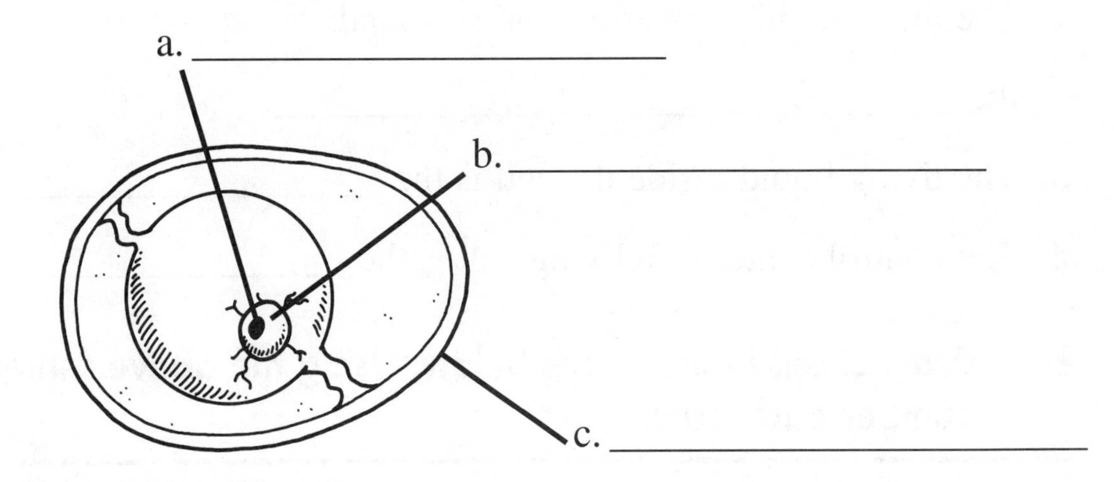

D. **Using your textbook and a dictionary, write the meanings of these science words.**

1. Cell membrane ______________________________

____

2. Cytoplasm ______________________________

3. Nucleus ______________________________

Name ______________________ Date ____________

What Does A Plant Cell Look Like?

A. **You will need a lettuce leaf, an onion ring, iodine, a microscope, a microscope slide, and a cover slip.**

B. **Peel the thick skin off the inside of a lettuce leaf. Put it on a microscope slide.**

C. **Place a drop of iodine on the skin and cover it with a cover slip.**

D. **Look at the lettuce skin through a microscope.**

1. What shape are the lettuce skin cells?

____

____

2. The nucleus will be either brown or yellow. What is it shaped like?

____

E. **Make a slide with the skin from the inside of an onion ring, the same way you made the lettuce slide. Look at the slide through the microscope.**

3. How are the onion skin cells and the lettuce cells alike? ________

____

F. **Look for some very tiny cells that are something like a doughnut. These cells have holes in the middle that open and close. The plant breathes through these cells.**

4. How many "breathing" cells did you find?

____

Name ______________________ Date ______________

Make a Model of an Animal Cell

You will need a jelly bean, some white clay, some blue clay, and a plastic knife. Put your jelly bean in the middle of a small ball of white clay. Then put a thin layer of blue clay all around your white ball. Now use your knife to cut through the ball. Be sure to cut through the jelly bean.

Observations

1. What does the white clay represent?

____

____

2. What does the blue clay represent?

____

____

3. What does the jelly bean represent?

____

____

Conclusions

4. How is the ball like a cell?

____

____

____

5. What are the main parts of a cell?

____

____

____

Name ________________________________ Date ______________

WHERE DO CELLS COME FROM?

Everything starts from just one cell. Did you ever wonder how whales, mosquitoes, mushrooms, and redwoods can all have so many? Imagine that you have a funny kind of cell called a "clock cell." Every hour each cell divides in two. Starting with one cell, there will be two cells in one hour, four cells in two hours, and so on.

How many cells do you think there will be in 12 hours? To find out, double the number of cells for each hour around the 12-hour clock.

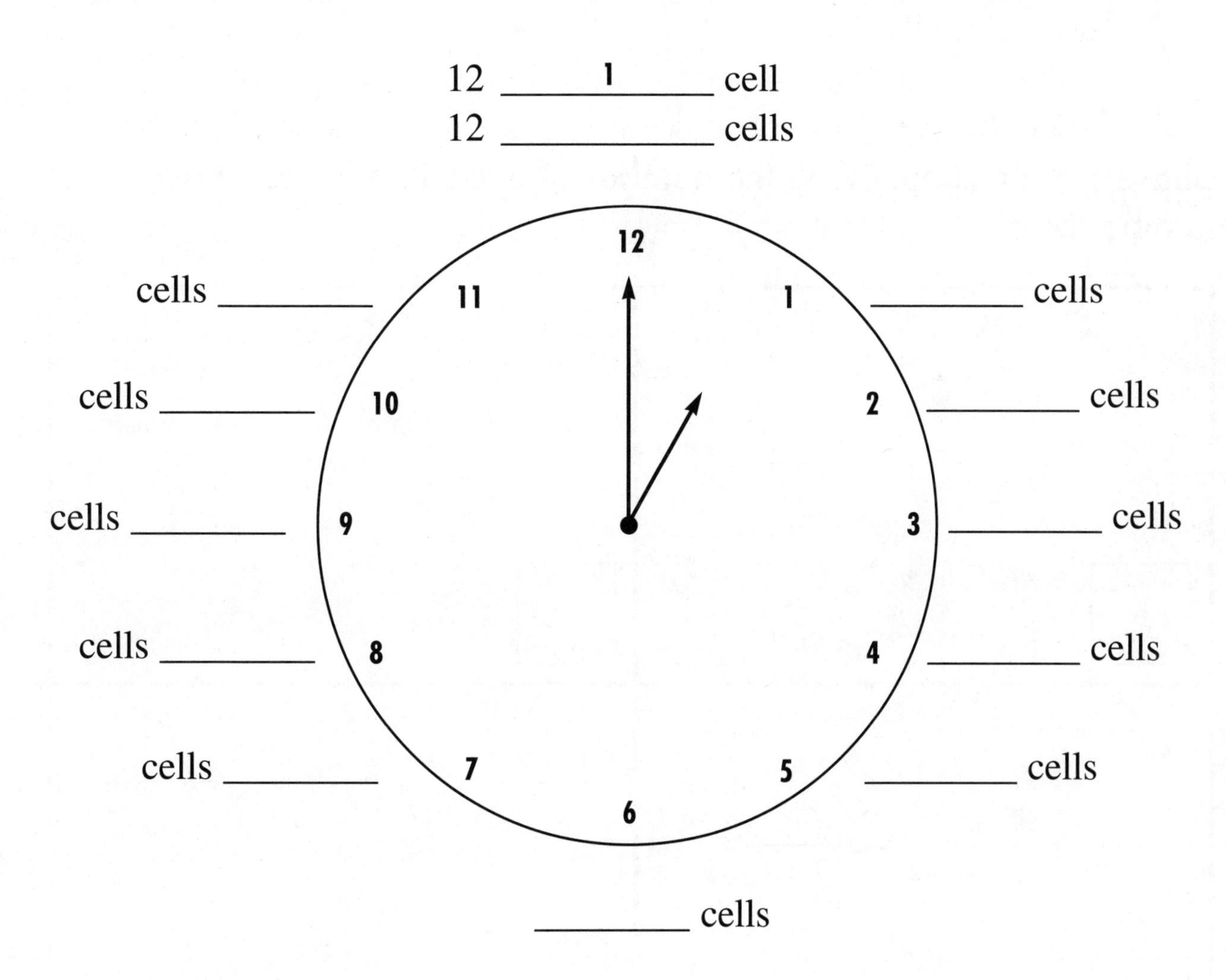

Name ______________________________ Date ______________

How Do Cells Divide?

Living things grow because their cells grow and divide, making more cells. The more cells an organism has, the larger it will be. How do cells divide?

1. First, the cell takes in extra food and becomes larger.
2. Next, the nucleus and the things inside the nucleus begin to divide.
3. Then, the cytoplasm divides.
4. Finally, the two parts of the parent cell pull apart to become two new cells.

In the boxes below, draw an animal cell during the four steps, or phases, of dividing. Write the number of each step in the small number box.

☐	☐
☐	☐

Name ________________________________ Date ________________

Cells Make More Cells

A. Read these sentences. Then put them in order to describe cell division.

______ The nucleus begins to divide.

______ The cell becomes larger.

______ The cell takes in extra food.

______ The two parts of the parent cell pull apart and become two new cells.

______ The cytoplasm divides.

Using the above sentences, write a paragraph about cell division.

____

____

____

____

____

B. Which of these living things takes the longest time to become an adult? Put a circle around the correct answer.

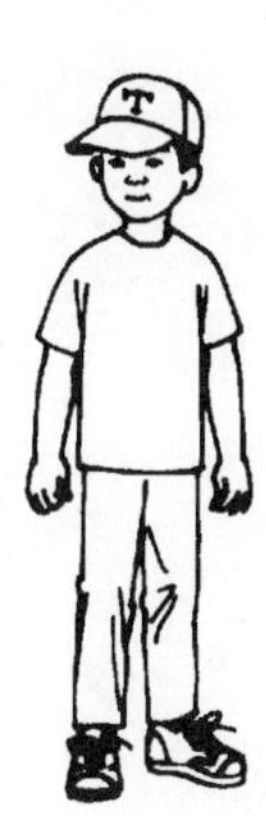

Name ______________________ Date ______________

Growing Organisms

Color these pictures, then cut them out. Put them in order to show how a frog develops.

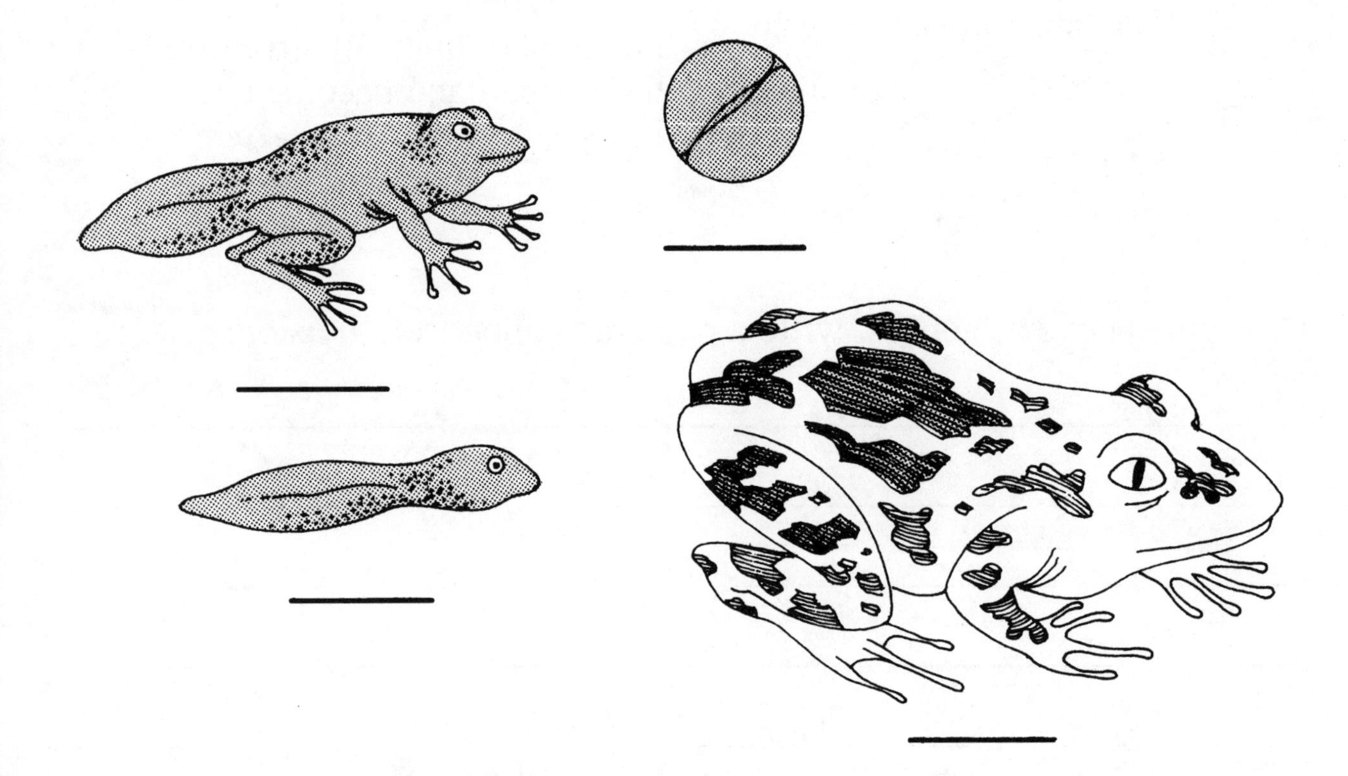

Name ______________________________ Date ______________

DIFFERENT TYPES OF CELLS

There are different types of cells that do different types of things in plants and animals. For example, there are many different types of cells in the human body. There are bone cells, blood cells, skin cells, muscle cells, and so on. Plants have different types of cells, too.

The box below has the names of different types of cells. Write each type of cell under the plant or animal below to tell if it is a plant or an animal cell.

muscle cells	cells with root hairs
bone cells	skin cells
cells with chloroplasts	blood cells
cells with cell walls	

______________________ ______________________

______________________ ______________________

______________________ ______________________

____

Name ______________________________ Date ______________

One-Celled Organisms

Some organisms are one-celled organisms. Two common one-celled organisms are *bacteria* and *protozoans.*

Scientists classify bacteria into three groups by shape. The groups are round, spiral, and rod-like.

Protozoans are classified in another way.

- There are some, called paramecium, that are covered with tiny hairs, called cilia, that help them move.
- There are some, called euglena, with thread-like whips that help them to move and to eat.
- There are others that move one part of their body and then drag along after it. They change shape constantly. They are amoebas.

Look at the pictures below. Three of them are bacteria, and three are protozoans. Put an X on the pictures of the bacteria. Put a circle around the protozoans.

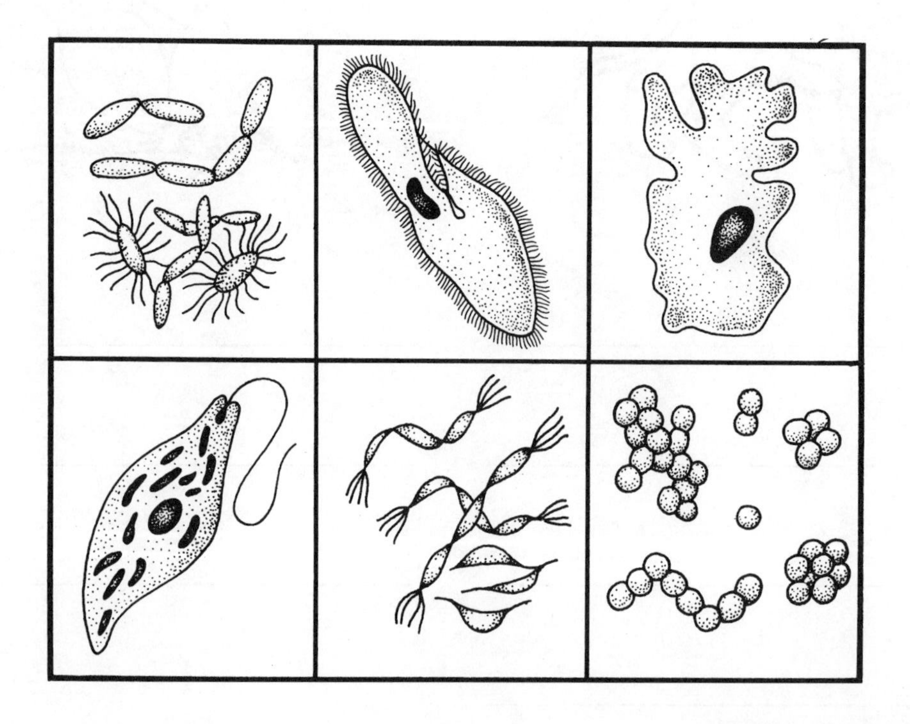

Name ______________________ Date ____________

WHERE DO YOU FIND ONE-CELLED ORGANISMS?

Pond water contains tiny animal-like creatures that have only one cell. You can grow some of these creatures and look at them under a microscope.

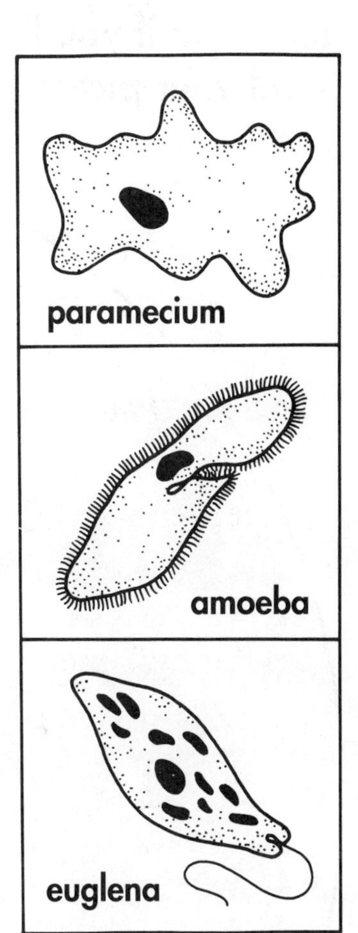

A. **You will need a glass jar, a microscope, a microscope slide, a cover slip, a soda straw, dry grass, and water from a pond or brook.**

B. **Fill your jar about half full of the pond or brook water. Put a handful of dry grass in the water.**

C. **Set the jar in a warm, dimly lit place for about 10 days.**

D. **Lower one end of the soda straw close to the bottom of the jar. Place your finger over the top of the straw to hold the water in. Use the straw to move a drop of water from the bottom of the jar to the slide.**

E. **Put the cover slip over the water drop and place the slide on the microscope. Look through the microscope. Move the slide until you see small creatures darting around.**

Answer these questions.

1. What do the creatures look like?

____

____

2. In what area do you find most of the creatures?

____

3. Why do you think they stay there?

____

Name ______________________________ Date ______________

DRAWING A ONE-CELLED ORGANISM

You are learning about the tiny organisms that can only be seen with a microscope. In this activity, you will draw a type of protozoan you might see if you looked at a drop of pond water under a microscope. Label your picture.

Name ______________________ Date ______________

THE LIVING CELL

Look at the drawing of the plant cell. The drawing shows what enters and leaves the cell as it makes its own food.

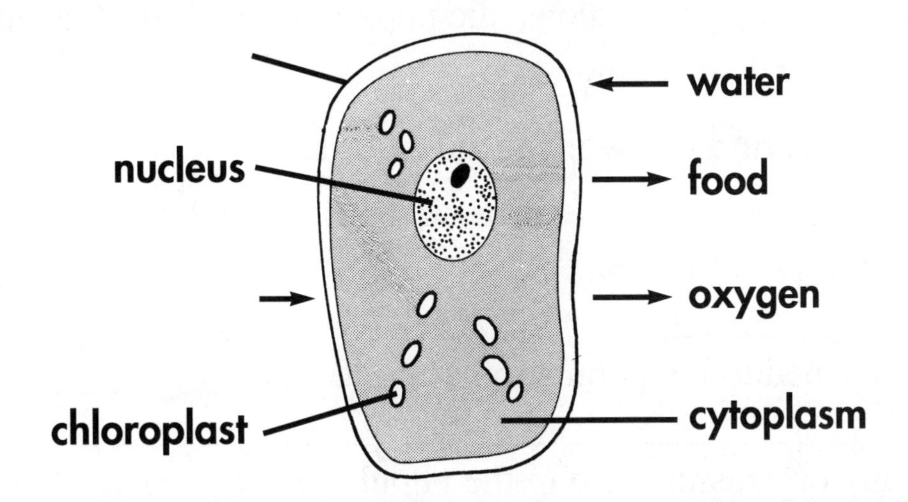

Now, write a paragraph that tells how the cell makes its own food.

____

____

____

____

____

____

____

____

____

Name ______________________ Date ______________

Groups of One-Celled Organisms

A number of one-celled organisms living in the same place is called a population. These organisms are a lot like people. All populations change just as our does. The graph shows the population growth of a simple organism compared with time.

Study the graph and answer the questions.

1. How many organisms were there on the first day? ______________

2. On which days did the number of organisms grow the most? ______________

____

3. What happened to the population after 20 days? ______________

____

4. How many organisms were in the population at

a. 10 days? ______________

b. 15 days? ______________

c. 20 days? ______________

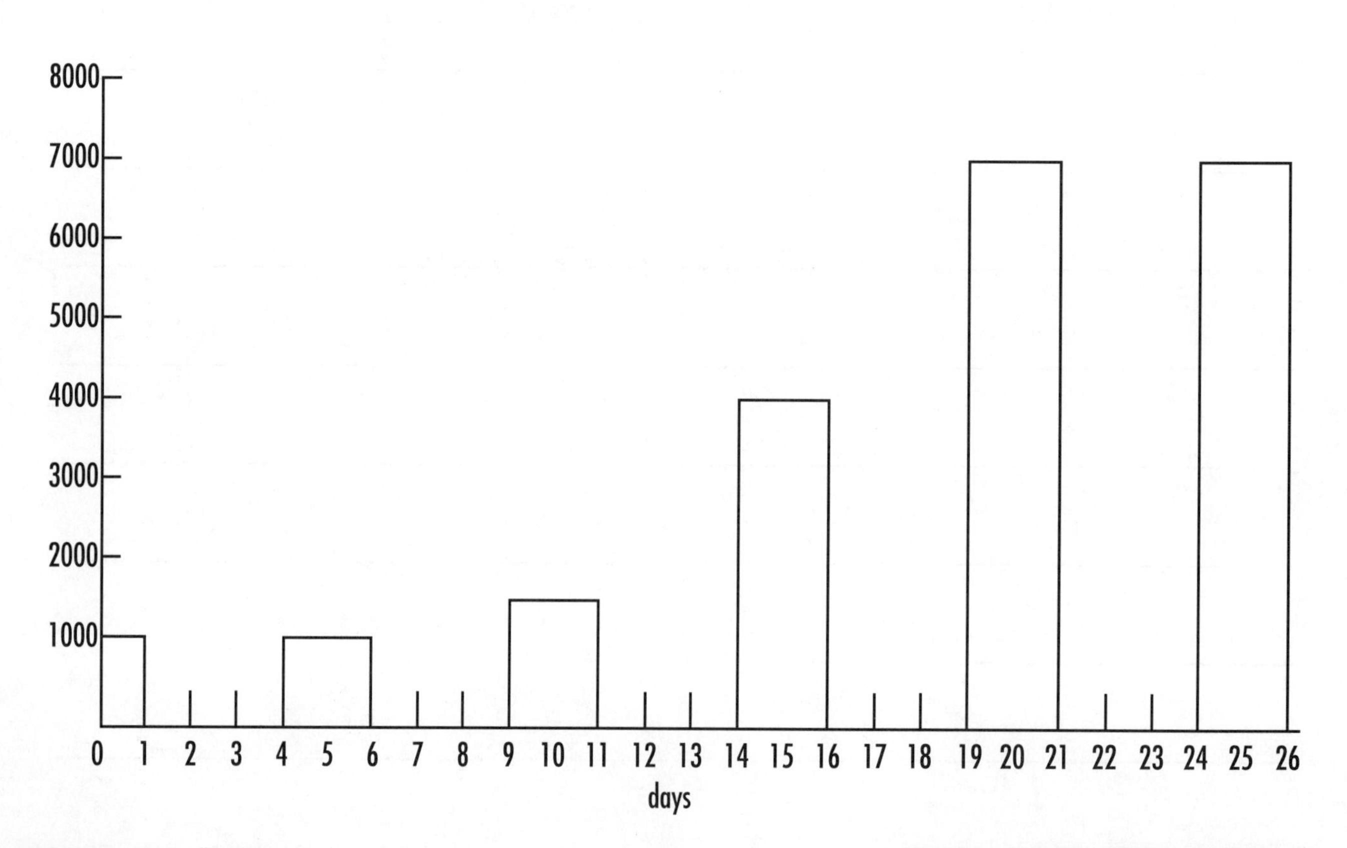

Name ______________________________ Date ______________

Simple Food Users

The words in the sentences below are out of order. Put the words in the right order so that they make sense. You will learn something about molds and other simple organisms. Be sure to begin each sentence with a capital letter.

1. that organism a simple food is spoils mold.

____

2. cloth grows on mold that is kind of a mildew.

____

3. a organism is bread to yeast used make simple.

____

4. soil turns mold and leaves into dead wood.

____

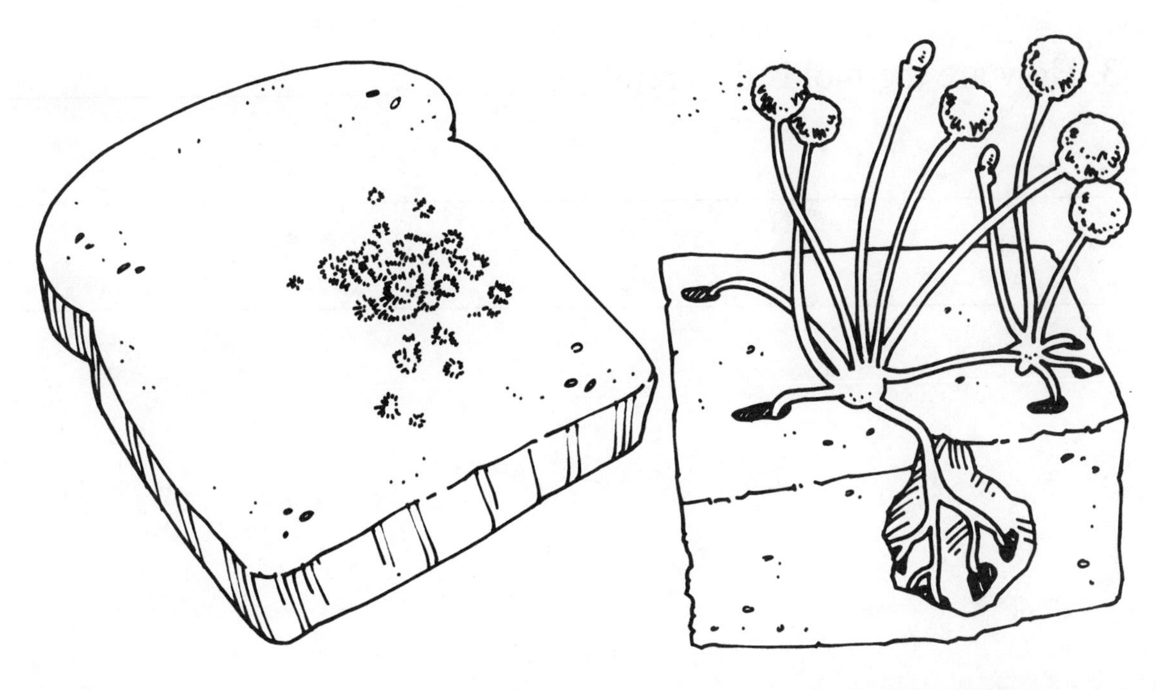

Name ______________________________ Date ______________

Where Does Mold Grow?

A. You will need a large jar with a lid, soil, an orange peel, a banana peel, cheese, and bread.

B. Lay the jar on its side. Put some clay under it so it will not roll. Spread some soil in the jar. (Note: Use soil from outdoors, not sterilized potting soil.)

C. Place a small piece of bread on the soil. Then add a piece of cheese, a piece of orange peel, and a piece of banana peel.

D. Put the jar in a dim, warm place. Look at it each day for two weeks.

Answer these questions.

1. On what food did the mold grow first?

____

2. Do different kinds of mold grow on different foods? ____________

3. How are the molds different? ______________________________

____

____

Name ______________________________ Date ______________

BUILDING YOUR SCIENCE VOCABULARY

Below are four puzzles. The answer to each is a kind of simple organism. Can you solve the puzzles?

Example

Note: This is not an example of a simple organism.

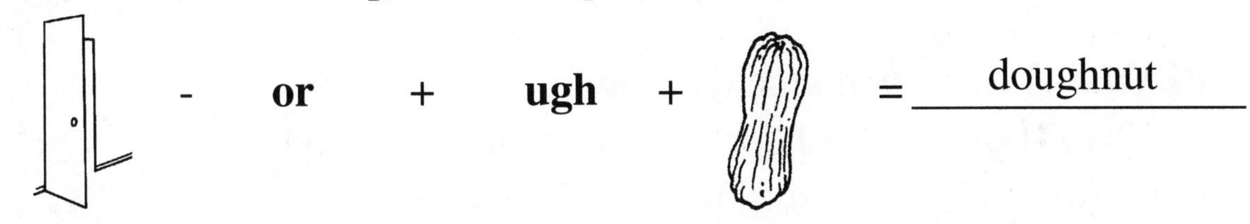

- or + ugh + = doughnut

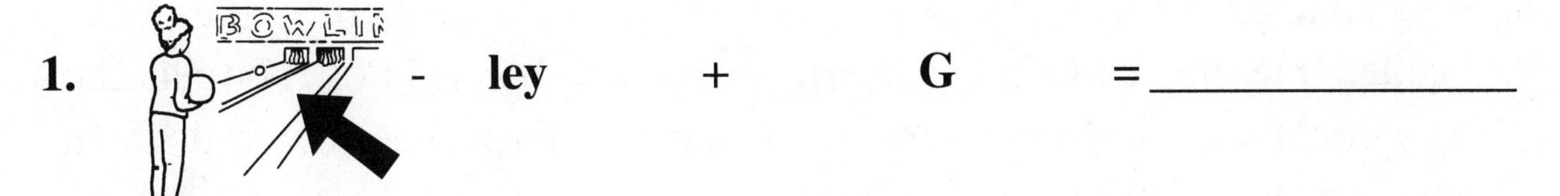

1. - **ley** + **G** = ______________

2. - **ELL** + N W E S = ______________

3. - **ney** + - **i** = ______________

4. - **l** + **dew** = ______________

Name ______________________________ Date ______________

Unit 1 Science Fair Ideas

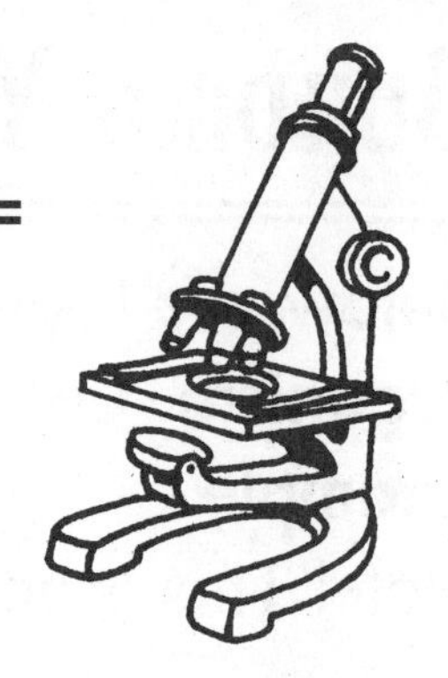

A science fair project can help you to understand the world around you better. Choose a topic that interests you. Then use the scientific method to develop your project. Here is an example:

1. **PROBLEM:** How do molds grow?
2. **HYPOTHESIS:** Mold does not make its own food to grow. It gets its food from things that are, or were, alive.
3. **EXPERIMENTATION:** Get several jars and put different things inside. One may contain bread, tomato, or another kind of food; another may contain a non-living thing such as rocks. Experiment with different amounts of moisture, air, and light.
4. **OBSERVATION:** The jars with more moisture and less light and which contain things that were once living, such as bread or other foods, grew mold more quickly.
5 **CONCLUSION:** Molds grow best in dark, damp places and live off things that were once living.
6. **COMPARISON:** Conclusion agrees with hypothesis.
7. **PRESENTATION:** Display the model with your project. Label and describe each jar.
8. **RESOURCES:** Tell of any reading you did to help you with your experiment. Tell who helped you to get materials or set up your experiment.

Other Project Ideas

1. How is food preserved, or kept fresh? Do the things added to food to keep them fresh longer work?
2. What kinds of single-celled organisms are found in a pond? In your home?
3. What is the difference between good and harmful bacteria?
4. What is the difference between a living and a non-living thing?

Unit 2: Plants
Background Information

Classification

The plant kingdom contains about 450,000 different kinds of plants, which are each classified into several divisions. The four main classifications for plants are:

algae (almost all live in water; from microscopic single-celled plants to seaweed),
bryophyta (mosses and liverworts; live in moist places; produce spores),
pteridophyta (ferns, clubmosses, horsetails; no flowers), and
spermatophyta (largest group—over 350,000 species; reproduce by way of seeds).

Spermatophytes are divided into two categories: the gymnosperms and the angiosperms. Gymnosperms, or "naked seed" plants, have seeds in cones, like pinecones from conifer trees. The angiosperms, or "covered seed" plants include all of the flowering plants.

Flowering plants are the most numerous type of plant on Earth. They are further classified into groups. Some of the common groups of flowering plants are:

grass family (corn, sugar, barley, rice, wheat),
lily family (violets, hyacinths, tulips, onions, asparagus),
palm family (coconut, date),
rose family (strawberries, peaches, cherries, apples, and other fruits),
legume family (peas, beans, peanuts),
beech family, and
composite family (sunflowers and others with flowers that are actually many small flowers).

Fungi are sometimes classified with plants, and sometimes they are not. A two-kingdom classification will include molds and fungi with plants, while a three-kingdom system will not. This is because molds and fungi lack chlorophyll and cannot produce their own food. They also lack roots, stems, and leaves, and they reproduce from spores that are distributed through the air or water.

Plants are also classified as vascular and nonvascular. Vascular plants have tubes that bring the liquids the plants need from their environment up through the stalk. The tubes also help to support the plants. Nonvascular plants, such as mosses, do not have tubes. They are shorter because they must remain close to their source of moisture. They get the water and nutrients they need through their root systems.

Photosynthesis

Most plants are green. The reason that green plants are green is because they contain chlorophyll, most of which is in the leaves. There are some plants that contain chlorophyll but whose leaves are not green. This is because the chlorophyll has been masked by other pigmentation in the plant. Chlorophyll is necessary for the making of food, but the chlorophyll itself is not used in the food that is made.

Photosynthesis depends on light. A plant that is deprived of light loses its chlorophyll (and its ability to make food) and eventually will die. Plants take in the energy from the sun and carbon, oxygen, and hydrogen from the air and water. They change these raw materials into carbohydrates and oxygen. The carbohydrates are used and stored in the plants for food. The oxygen is released into the air and water where the plants live. In this way, plants constantly replenish the Earth's oxygen supply.

Mushrooms and molds are not green because they do not have chlorophyll. They cannot make their own food. Molds and mushrooms depend on other organisms for their food. Mushrooms and molds are fungi.

Reproduction

Plants reproduce from seeds in flowers, from seeds in cones, or from spores. The seeds form after fertilization of their egg cells by male cells from pollen grains. Pollen can be carried to the egg cells by bees or other insects, by the wind, or by animals.

Seeds contain tiny plants called embryos around which a store of food is packed. In some seeds, such as bean seeds, the food is stored inside the embryo. Seeds are spread by animals and the wind. When the seeds in a cone are ripe, the cone will open, and the seeds will float to the ground or be carried by the wind. Some seeds have tiny parachutes to help them drift. A seed needs moisture, warmth, and oxygen to begin growing into a new plant. If conditions are not right for germination, some seeds can remain in a resting state for hundreds of years.

Pollination

The first thing people usually notice about a flower are its petals. The petals of a flower are brightly colored and designed to attract the insects that will help to pollinate it. When an insect comes to a flower to get nectar for itself, its body may touch the stamen of the flower. The stamen is covered with pollen that it has produced. The pollen is carried on the insect's body to the next flower. The pollen goes down the pollen tube in the pistil of the new flower. At the bottom of the pistil are the ovules, or the tiny beads that will grow into seeds. If the pollen is from the right type of flower, the ovules are fertilized and begin to grow. Some flowers can self-pollinate, but many prevent this from happening because their stigmas and stamens ripen at different times. Flowers that depend upon insects to carry their pollen are cross-pollinated.

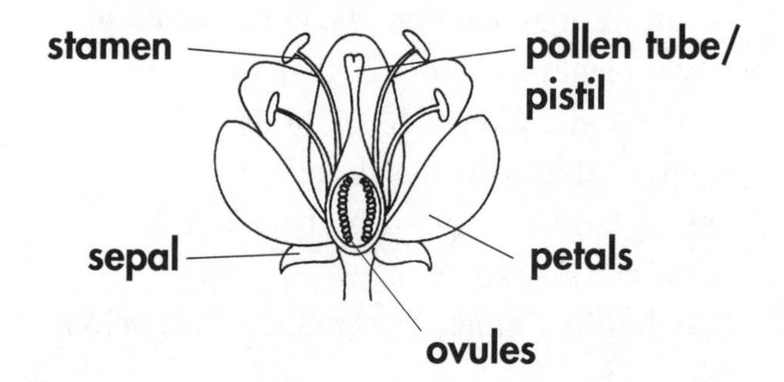

Life Cycle

The life cycle of a seed plant begins with an embryo. An embryo is an undeveloped living thing that comes from a fertilized egg. The eggs in a flowering plant are called ovules. When the ovules are fertilized, they begin to grow. A seed is a complete embryo plant surrounded by the food it needs to grow and protected by a coating. When the seed is planted, or lands on the ground, it begins to sprout. It grows into a seedling, then an adult plant that develops flowers in which new seeds grow.

Name ______________________________ Date ______________

GREEN PLANTS

Plants need oxygen, water, and light to grow. What would happen if one of these things were taken away from a plant?

1. Cut out a small circle of construction paper. Use a paper clip to attach the circle to a large leaf of a plant. Wait one week. When you remove the circle, what do you see?

   ____

   Why did this happen? ______________________________

   ____

2. Put a plastic bag around a small plant and close the bag with a twist tie. Leave the plant for several days. What happens to the plant?

   ____

   ____

   Why did this happen? ______________________________

   ____

3. What do you think would happen to a plant if it could not get any water?______________________________________

   Why would this happen?______________________________

   ____

4. Plants need ______________, ______________, and ______________ to grow.

Name ______________________ Date ______________

Tubes or No Tubes?

Some plants get the water they need through tubes in their stems. Some plants do not have tubes. How do they get water? To find out, compare a celery stalk with a mushroom. **You will need:**

red or blue food coloring

paper cups	**water**
toothpicks	**celery stalk**
knife	**mushroom**
hand lens	

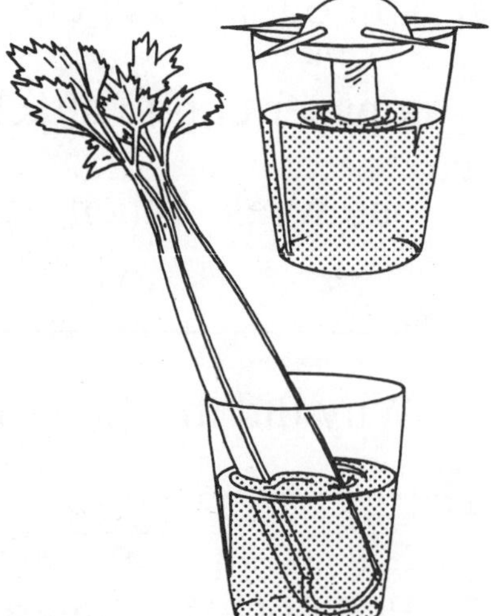

1. **Fill both cups 2/3 full with water. Add a few drops of food coloring to each cup. Cut a thin slice off each stem.**
2. **Put the celery stalk in one of the cups. Put four toothpicks into the cap of the mushroom. Then put it into the other cup. Leave them in the water overnight.**
3. **The next day, cut the stems in half. Observe them with a hand lens.**

Fill in the chart.

	Celery	**Mushroom**
Does the stem have spots of colored water in it?		
Where is the water in the stem?		
Does the plant have tubes?		

Name ______________________ Date ______________

Tree Rings

The trunk of a tree is made mostly of the tubes that carry water. When a tree is cut down, you can see rings of tubes in the tree stump. Each ring shows one year of growth. During a year of fast growth, a tree makes a wide ring. During a year of slow growth, a tree makes a narrow ring.

If you can find a stump, count the rings. They will tell you how old the tree was when it was cut down. For example, if you count 50 rings, then the tree lived for 50 years.

Some trees grow faster than others. Some trees live longer than others. Those that live longest and grow fastest have the biggest trunks. The graph shows how big five different kinds of trees can grow. The measurement used to compare them is the diameter. Diameter is the distance through the center of the trunk.

Look at the bar graph. Then answer the questions on page 42.

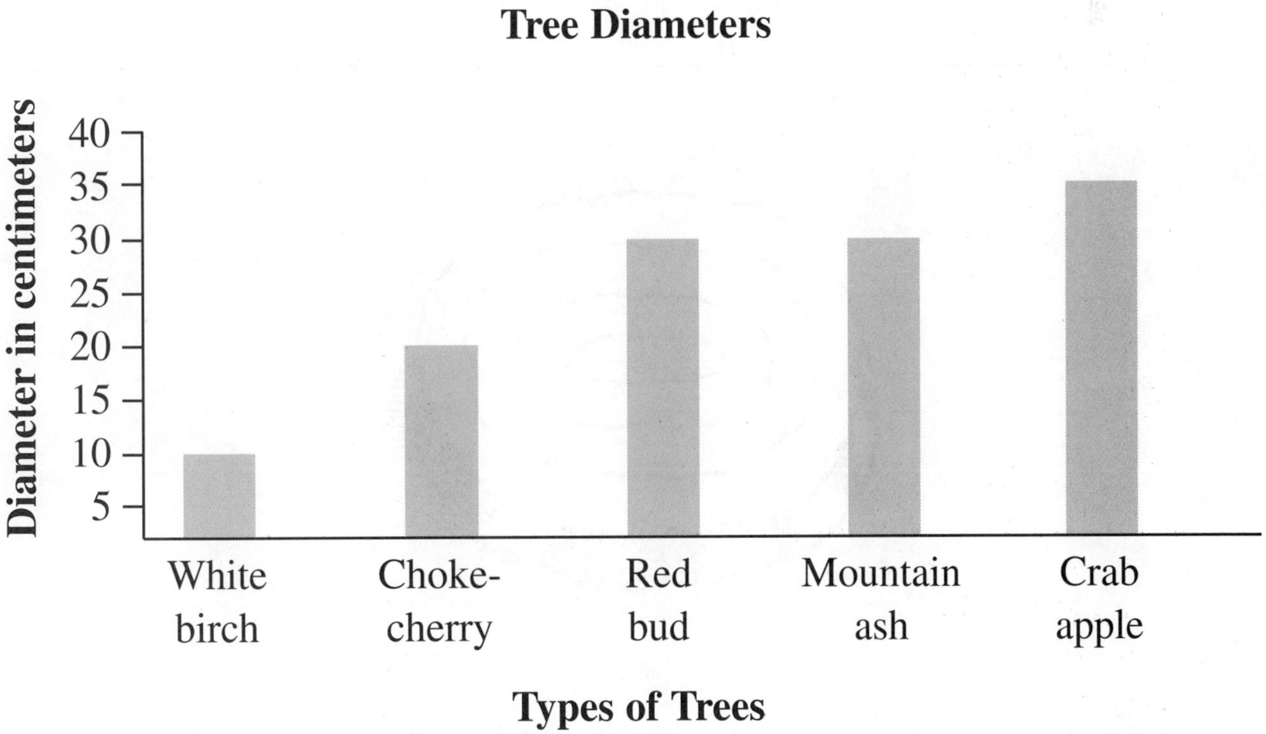

Tree Diameters

Diameter in centimeters

40
35
30
25
20
15
10
5

White birch
Choke-cherry
Red bud
Mountain ash
Crab apple

Types of Trees

Go on to the next page.

Name ______________________ Date ____________

TREE RINGS, P. 2

1. Which fully grown tree has the smallest trunk? ____________

____

2. Which two trees can grow to have the same diameter? ____________

____

3. A fully grown chokecherry has a diameter of ____________

____.

4. Which tree can grow the largest trunk? ____________

____

5. A mountain ash probably grows faster or lives longer than

a ____________ and ____________________.

Name ______________________ Date ____________

OBSERVING LEAVES

What happens to a leaf of a plant if it does not get sunlight? What happens to the plant?

To find out, you will need:

geranium plant | **tape**
aluminum foil | **sunny window**
water

1. **Look at the leaves of your plant. Fill in part A of the chart.**
2. **Cover both sides of one leaf with aluminum foil. Make sure no light can get to the leaf. Use tape to hold the foil in place.**
3. **Put the plant in sunlight. Water it whenever the soil begins to dry out.**
4. **After 10 days, remove the aluminum foil. Compare the covered leaf with the uncovered leaves. Fill in part B and C of the chart. Then answer the questions.**

Changes in Geranium Leaves

	My Observations
A. leaves at the start	
B. uncovered leaves after 10 days	
C. covered leaves after 10 days	

Go on to the next page.

Name ______________________ Date ____________

Observing Leaves, p. 2

Answer these questions.

1. What happened to the covered leaf?

____

____

____

2. Did the uncovered leaves change? ______________________

Why? Why not? ______________________

____

3. What might happen if all the leaves were covered? ______________________

____

____

____

Name ______________________ Date ______________

COMPARING LEAVES

A. Are leaves from different plants alike? To find out, collect two different green leaves. Tell about them in the chart below. Measure the leaves with a ruler. If you cannot describe them easily, draw pictures. Show the tip and edge of each leaf.

	Leaf A	Leaf B
Color		
Length		
Width		
Shape of tip		
Shape of edge		

Answer these questions.

1. How are the leaves the same? ______________________

____

____

2. How are they different? ______________________

____

____

Go on to the next page.

Name ______________________________ Date ______________

Comparing Leaves, p. 2

Plants are green because they contain chlorophyll. Chlorophyll helps them make food. Without chlorophyll, plants cannot make food.

B. You can write your name with the coloring in plant leaves.
You will need:
2 leaves pencil with dull point

1. Place a leaf on the bottom of this paper. Press the pencil in to the leaf to make dots on the paper. Try to make letters in this way.

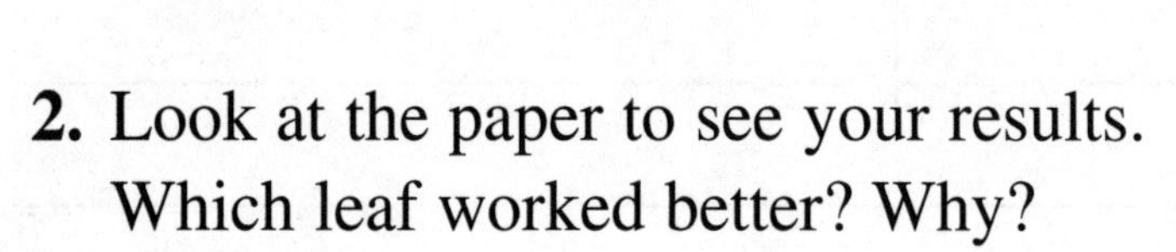

2. Look at the paper to see your results. Which leaf worked better? Why?

____

____

3. What gives the leaf its green color?

____

Name ______________________________ Date ________________

Can the Leaves of a Plant Change Direction?

Place a small plant by a window. Be sure the leaves are facing toward you and away from the window. Draw a picture of your plant. After one week, draw another picture of your plant. Write a sentence to tell what happened.

Week 1	**Week 2**

What happened?

____

____

Name ______________________________ Date ______________

How Does Sunlight Affect Water Plants?

To answer this question, you will need:
2 small glass jars of water (let stand overnight)
Water plant (elodea recommended), 2 pcs

Place one elodea plant in each jar. Place one jar in a dark place, like a closet. Place the other plant on a windowsill where it will get sunlight. Look at the plants after one hour.

Observations

1. How did the plants change?

____

____

____

____

____

2. Which plant has bubbles coming from it?

____

____

Conclusion

3. How does sunlight affect water plants?

____

____

____

____

____

Name ______________________________ Date ______________

Inside a Flower

Have you ever seen the inside of a flower? If you were a bee, you would have seen the inside of many flowers! Bees help flowers to reproduce, or make more flowers. A bee goes to a flower for food to eat. It gets pollen on its body from the **stamen** of the flower. Then the bee goes to the next flower. The bee pollinates the flower. The pollen from the bee's body goes down the **pollen tube** in the **pistil** of the new flower. The pollen fertilizes the **ovules**, or tiny seeds, that are at the bottom of the pistil. The seeds begin to grow.

Below is a picture of the inside of a flower. See if you can name the parts of the flower. Use an encyclopedia to help you.

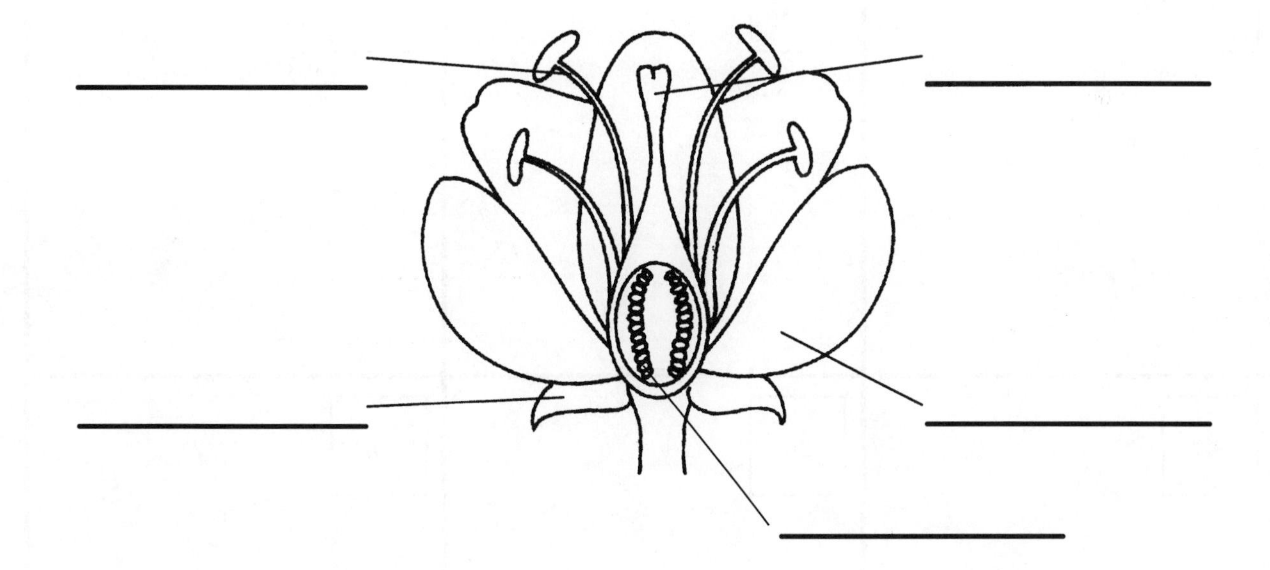

Name ______________________________ Date ______________

Life Cycle of a Plant

Living things have life cycles.

1. The life cycle of a plant begins with an embryo. An embryo comes from a fertilized egg. The egg cells of a seed plant develop in its flowers.
2. The seed sprouts.
3. A seedling grows.
4. The plant develops into an adult plant.
5. The adult plant produces flowers.
6. Inside the flower, new egg cells develop.

In the boxes below, draw pictures of the life cycle of a plant. You may use an encyclopedia to help you.

1.	2.	3.
4.	5.	6.

Name ______________________________ Date ________________

FINDING THE EMBRYO IN A BEAN SEED

Examine a bean seed. Remove the seed coat. Separate the two halves of the bean with your fingernail. Can you find the embryo? Draw a picture. Show what you found in the bean seed. Show where the embryo was found.

Name ______________________________ Date ______________

Designing with Seeds

Seeds come in different shapes, sizes, and colors. These differences can help you make a seed design. **You will need:**

several kinds of seeds
envelopes
sheet of cardboard
pencil
glue

1. Collect different kinds of seeds. A handful of each should be enough. Put each kind of seed in a separate envelope. Label each envelope with the name of the seed. From the kitchen you may be able to get rice, fruit seeds, and several kinds of beans. Outside, you may be able to find acorns, burrs, or grass seeds.

2. In a chart, write the names of the seeds you have found. Then look at each group of seeds. What color are they? Are they round, oval, square, or some other shape? Are they smooth, fuzzy, rough, or some other texture? Describe each type of seed in the chart.

3. Now draw a design on a sheet of cardboard. Paste your seeds on the cardboard to fill in the design.

My Seed Collection

Name of Seed	Color	Shape	Texture

Name ______________________ Date ______________

Beds for Mushrooms

Read the story about mushrooms. Then follow the directions.

What grows in old mines, deep caves, and dark cellars? Mushrooms do!

Mushrooms are fungi. They do not have chlorophyll. They do not need sunlight. They do not make their own food. Instead, they get their food from dead plants and animal material.

To grow mushrooms, a farmer sprinkles spores onto a bed of grain. The spores grow into threads. The farmer spreads the threads over large growing beds. Then the farmer covers them with a layer of fine soil.

These large growing beds are filled with straw. The beds cannot get too wet or too cold. They must be just right or the mushrooms won't grow.

About six weeks after the threads are planted, the first mushrooms appear. They spring out of the soil in groups. All the mushrooms in a group are picked at the same time. They are then sorted by size. Big ones are sold fresh to food stores. Small ones are canned.

The threads in a good mushroom bed will grow for about five months. Then no more mushrooms appear. The farmer destroys the bed and makes a new one.

Go on to the next page.

Name ______________________________ Date ______________

Beds for Mushrooms, p. 2

Circle the right word or words to complete each sentence.

1. Mushrooms can be grown in (greenhouses, caves).

2. Mushrooms (do, do not) contain chlorophyll.

3. Mushrooms (get food from dead plant and animal material, make food from sunlight).

4. Mushroom beds must not get (too wet or too dry, too dark or too light).

5. A mushroom bed can often grow mushrooms for (six weeks, five months).

6. When a mushroom bed stops growing, the farmer (destroys it, spreads new spores on it).

7. Mushrooms (can, cannot) be grown in dark cellars.

Name ________________________________ Date ________________

Growing Cress on Your Windowsill

You can grow salad cress seeds on your windowsill. Mustard seeds also work well. **You will need:**

an egg carton paper towels water brown paper seeds

A. Cut the top off the egg carton. Put several layers of paper towels into the top.

B. Wet the towels and sprinkle the seeds on them.

C. Cover the seeds with brown paper, and keep the paper damp.

D. When the seeds have sprouted, put them in a sunny place. Keep them moist.

E. Harvest the plants when they are 2 inches high. Cut them off with scissors near the roots.

F. They are delicious on sandwiches. Enjoy!

Name ______________________ Date ____________

Kitchen Garden

You can grow a garden without buying seed packages. Just look in your kitchen for seeds. You will probably find many that will grow.

To make a kitchen garden, you will need:

soil **scissors** **pencil** **milk cartons**

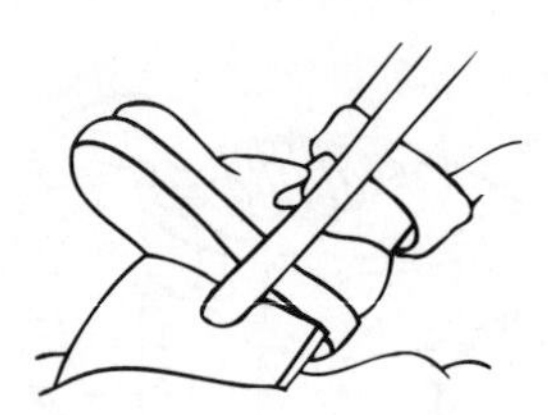

1. Look for seeds in your kitchen. You may find dried beans or peas, tomato seeds, pits from ripe oranges, grapefruits, or lemons.
2. Soak the beans and peas in water overnight. The other seeds can be planted right away.
3. Cut off the tops of the milk cartons. Add soil almost to the top. Use one carton for each kind of seed.
4. In each carton, plant one large pit or several small seeds. Label each carton with the name of the seeds or pit you planted.
5. Place your cartons in a warm spot. Keep the soil moist. When shoots appear, make sure they get plenty of sunlight.

Fill in the chart.

What did you plant?	How many days before the shoots appeared?
____________	____________
____________	____________
____________	____________

Name ______________________ Date ______________

Leaf Art

Water is drawn up through many plants through tubes in their stem. You can see the tubes in some leaves. They are the veins of the leaf.

Collect some leaves. Choose a leaf with lines, or veins, that you can see clearly. Place the leaf facedown on your desk. Place this paper over the leaf. Gently rub a crayon over the entire leaf. Can you see the veins now? Do rubbings of different leaves. Are all the patterns the same?

Name ______________________ Date ______________

Unit 2 Science Fair Ideas

A science fair project can help you to understand the world around you better. Choose a topic that interests you. Then use the scientific method to develop your project. Here is an example:

1. **PROBLEM:** How do different levels of light affect the growth of a plant?
2. **HYPOTHESIS:** Plants with the most light will show the best growth.
3. **EXPERIMENTATION:** Use several different small pots to plant the same kind of seed. A lima bean seed is good to work with. Number your pots and be sure to keep careful records of which pots are getting the most light, the least light, etc. Use two or three pots for each level of light to be sure your observations are accurate. Be sure to water your seeds lightly and regularly as they grow. The soil, moisture, and temperature of all your plants should be the same so that the only thing that is different is the lighting.
4. **OBSERVATION:** The plants that are allowed plenty of sunlight show the best growth. The plants with no sunlight do not grow well, if at all.
5. **CONCLUSION:** Plants need plenty of sunlight to thrive.
6. **COMPARISON:** Conclusion agrees with hypothesis.
7. **PRESENTATION:** Display all of your plants and your records of their growth. Try to display them in the manner in which you grew them. For example, if one of your plants was in total darkness, cover it with a labeled box.
8. **RESOURCES:** Tell of any reading you did to help you with your experiment. Tell who helped you to get materials or set up your experiment.

Other Project Ideas

1. What is the difference between a fruit and a vegetable?
2. What can change the color of a plant?
3. How does a seed grow into a plant?

Unit 3: Animals
Background Information

Classification

The animal kingdom can be classified into two large groups: the vertebrates (those with backbones) and the invertebrates (those without backbones). The backbone supports the body and provides flexibility.

The spinal cord extends from the brain through the backbone, or spine. Individual nerves branch out from the spinal cord to different parts of the body. Messages from the brain are sent throughout the body through the spinal cord.

Some animals without backbones are sponges, jellyfish, clams, worms, insects, and spiders. Some of these animals have networks of nerves throughout their bodies with no central nerve cords. Many, like insects, have hard exoskeletons that protect their bodies and give them shape.

Animals Are Suited to Their Environments

Animals live in almost every type of environment on Earth. Each kind of animal has become well suited to its environment through generations of adaptation. Those animals that are not suited to the environment, or that are poorly adapted, do not survive. The animals that are most fit for their environments continue to reproduce and make others like themselves.

Every part of an animal helps it to live in its particular environment. Some animals are colored in ways that help them to blend in to their environments. They are camouflaged to protect them from their enemies. Other animals are brightly colored to attract mates and help them with the continuation of their species. Animals' mouths and teeth are adapted to the types of food that they eat. Meat-eating animals have sharp teeth for tearing and ripping their prey, and other teeth for chewing the meat. Animals that eat leaves and grasses have large flat teeth for chewing.

The beaks and feet of different birds vary greatly. Some birds have thick, short beaks for cracking seeds. Others have long, slender beaks that they dip into flowers to reach nectar. The heron uses its long beak like tweezers to pluck fish from the water. Woodpeckers have beaks strong enough to hammer holes into trees as they look for insects. Eagles have strong, thick claws that enable them to grasp and carry their prey. The webbed feet of ducks help them to swim. Some birds have feet that help them climb, and others have feet that are best suited for perching.

Insects also have different types of mouth parts depending on their diets; some have chewing mouth parts to eat plants, and others have sucking mouth parts to get liquids.

Most animals are suited to either land or water life. An obvious adaptation for fish is the gills that allow them to breathe in the water. Lungs allow land animals to breathe in air.

Many animals protect themselves with camouflage. Other ways that animals protect themselves are by unique physical defense systems. Skunks emit foul-smelling odors when they are frightened. Porcupines bristle with barbed quills that will embed in the nose of any enemy that gets too close. Other animals have warning systems to alert the rest of a group to danger. Prairie dogs post sentries that will emit a loud whistling sound when danger approaches, and all the prairie dogs will go into their burrows immediately. Many insects have coloring that fools prey. The wings of some moths and butterflies look like large eyes. Predators believe that the insect is larger than it is, and leave it alone. Some animals not only mimic the color of their surroundings, but the shape, as well. A walking stick looks just like the sticks and bark among which it makes its home. Yet other animals will change color to fit their changing environments.

Adaptations to Environment

In order for living things to remain alive, they must respond to changes or conditions in the environment. Environments include all the conditions in which a living thing exists. This includes the food the organism needs, water, soil, air, temperature, and climate. Common environments are deserts, grasslands, and forests. Each of these larger environments contains many smaller environments. There are ponds and swamps in the forest. Deserts can be hot or cold. The living and non-living things in each environment interact with each other to survive. When environments are threatened or changed in drastic ways, the living things in the environment are also threatened.

Animals are adapted to their environments through structures and behaviors. The structures include the physical makeup of animals, some of which were mentioned in the previous section. The behaviors of animals include things like migration and hibernation. In winter, many birds migrate to warmer climates in the south. Some animals, like moose and caribou, also have migratory routes. Many animals hibernate, or sleep, through the winter months. They work through the fall to store food in their bodies that carries them through the winter months. While they sleep, their body processes slow.

Reproduction

All living things have a life cycle within which they take in food and gases, metabolize, excrete waste, reproduce, and die. If living things fail to repdroduce or to create healthy offspring, their species will die out. The California condor is one animal whose offspring have failed to thrive. So few of their chicks have survived in recent years that they may be in danger of extinction. It has been found that pesticides concentrated in the bodies of the adult condors have interfered with their reproductive abilities.

Animals reproduce in different ways. Some lay eggs, and others give birth to live young. Some offspring look like their parents while others do not. Most reptiles, amphibians, fish, and insects lay eggs. The young of many of these animals can move about and find food for themselves soon after they hatch. Birds also lay eggs, but the adult birds remain with the eggs and care for the young until they can find their own food. Most mammals bear live young. The young are fed milk from the body. Mammals spend more time than other animals feeding, protecting, and teaching their young to survive on their own. Animals that give birth to live young have fewer offspring than those that do not tend to their young. The young of human beings require more care from their parents than any other animal.

Metamorphosis

The life cycles of some animals include a metamorphosis. A metamorphosis is a complete change in the appearance of an animal. The most striking metamorphosis is the change from caterpillar to butterfly. Metamorphosis is controlled by hormones in the body. When the hormone supply keeping a caterpillar a juvenile stops, the caterpillar begins to become a chrysalis, or pupa. In a frog, the change is controlled by the thyroid gland. Crabs also undergo metamorphosis, and earwigs and grasshoppers undergo incomplete metamorphosis.

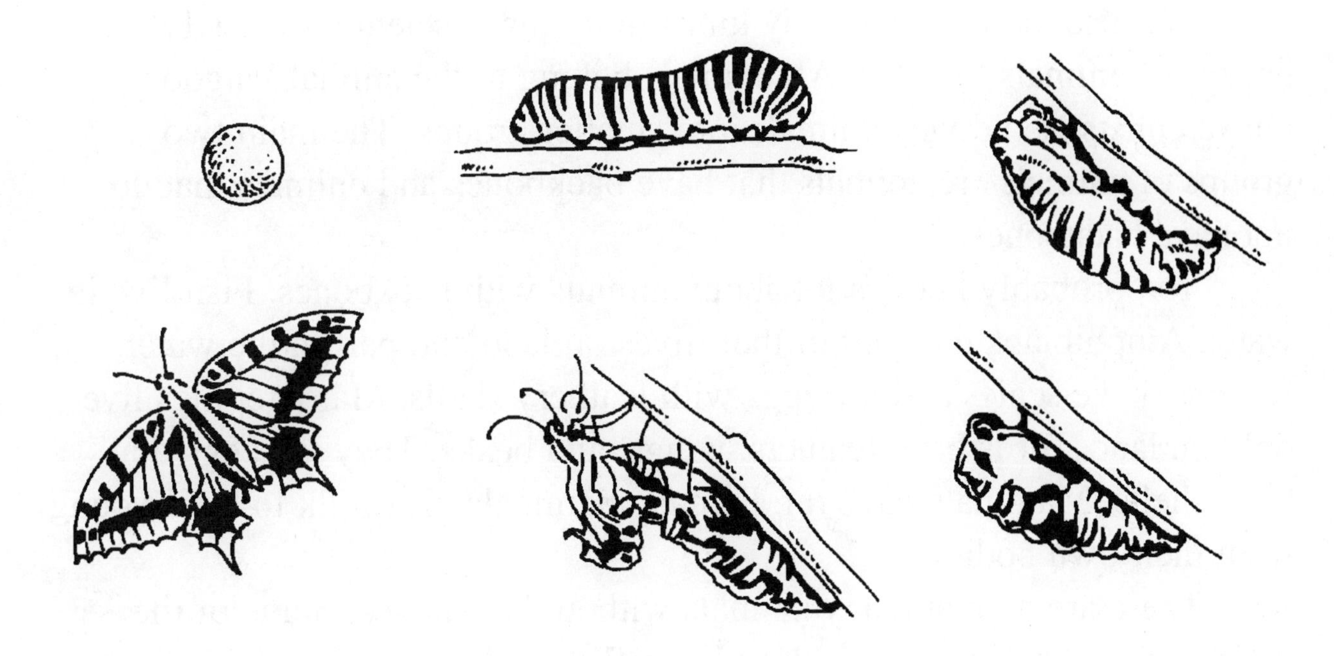

Name ______________________________ Date ______________

Which Group?

Read the summary, and then make a list of animals. Your list should have 5 groups: fish, amphibians, reptiles, birds, and mammals. Include as many examples under each group as you can think of.

What do rabbits, wolves, mice, cats, dogs, and humans all have in common? One of the things they all have is fur. Features that living things have are called traits.

Scientists use traits to separate living things into groups. Making groups of things is called classifying. When you classify things, you use

features of the things to group them. For example, if you have a pile of marbles, and you make groups of red marbles, green marbles, and blue marbles, you are classifying the marbles by color.

Scientists classify living things into five large groups called kingdoms. The kingdoms you probably know the most about are the plant kingdom and the animal kingdom.

Because there are so many kinds of animals, scientists classify groups of animals together. All animals belong to the animal kingdom, but scientists also divide animals into smaller groups. The main two groups of animals are animals that have backbones and animals that do not have backbones.

You probably know a lot about animals with backbones. Fish live in water. Amphibians live part of their lives on land and part in the water. Reptiles have scales and lay eggs with leathery shells. Many reptiles live only on land. Birds have feathers, wings, and beaks. They lay eggs that have shells. Mammals have fur. Female mammals give milk to their young from their own bodies.

There are a great many animals without backbones. Some of the most common are insects, such as butterflies and bees. Animals without backbones can be found nearly everywhere you look.

Name ______________________________ Date ________________

Animal Groups

A. Look at the pictures. In each group, circle the animal that does *not* have a backbone.

1. These animals can swim.

2. These animals can fly.

3. These animals lay eggs.

 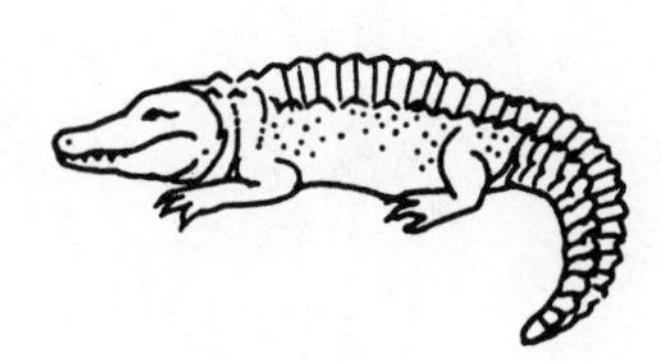 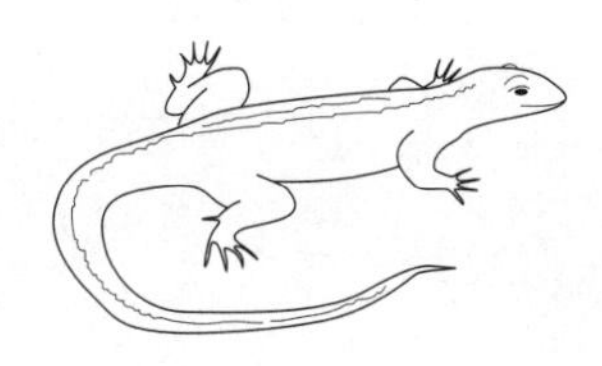 

4. These animals hop.

 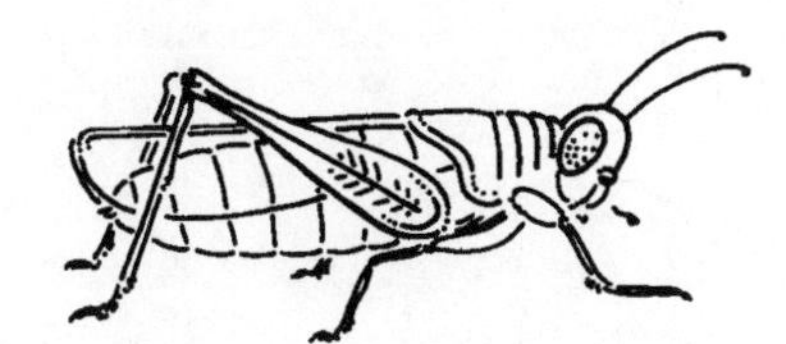 

Go on to the next page.

Name ______________________ Date ____________

ANIMAL GROUPS, P. 2

5. These animals can live in the ocean.

6. These animals live in a tree.

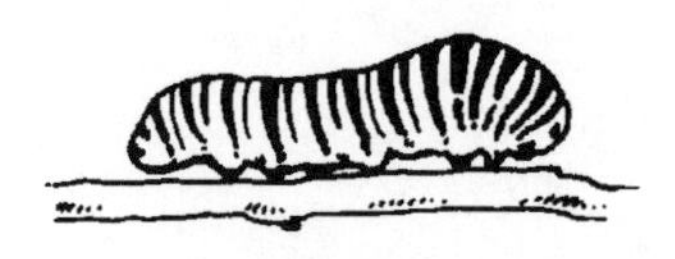

B. Look at the pictures of the animals you did not circle.

1. Color each bird.
2. Underline two reptiles.
3. Draw an X on two mammals.

Name ______________________ Date ______________

Classifying/Ordering

When you classify objects, you put them into groups according to how they are alike. Ordering is putting things into an order. For example, you might order things from first to last, smallest to biggest, or lightest to heaviest.

Think About Classifying/Ordering

Ask yourself:

1. What am I classifying?
2. What feature will I use to group my objects? Is there more than one way I can group my objects?
3. How does grouping my objects help me understand more about them?

Try It

The pictures below show different animals. The graphic organizer on the next page shows how the animals can be classified into three groups: those that live on land, those that live in water, and those that can live both on land and in water.

Go on to the next page.

Name ______________________________ Date ______________

CLASSIFYING/ORDERING, P. 2

1. Use the graphic organizer to classify the animals based on traits other than where they live. Think of a way to separate each group of animals into two additional groups. Write the names of the animals in the boxes. On the line under each box, write in the trait the animals in the box have in common. The first one has been started for you.

Animals
flamingo, parrot, tiger, dolphin, gecko, boa constrictor, guppy, shark, turtle, frog, wolf, water snake

Common Trait: Lives on Land

flamingo, parrot, tiger, gecko, boa constrictor, wolf

Common Trait: Lives in Water

dolphin, guppy, shark

Common Trait: Lives on Land and in Water

turtle, frog, water snake

tiger
wolf

Common Trait: fur

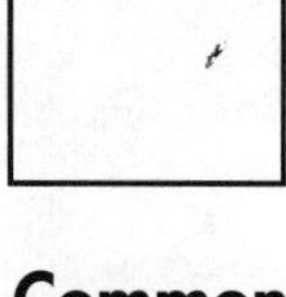

Common Trait: ____________

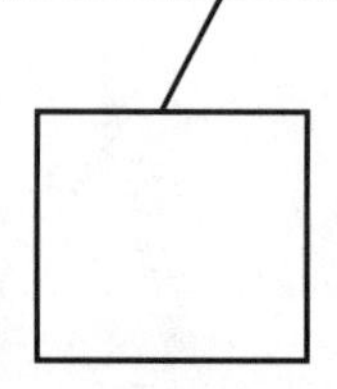

Common Trait: ____________

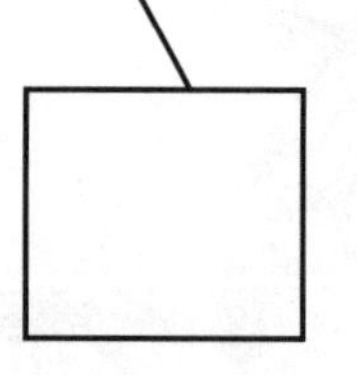

Common Trait: ____________

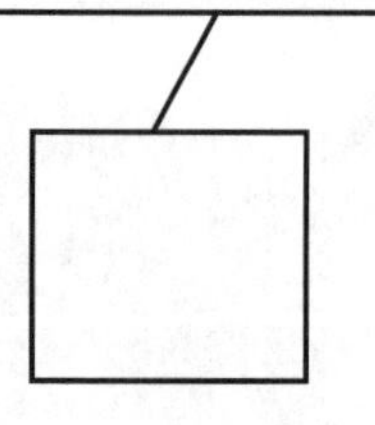

Common Trait: ____________

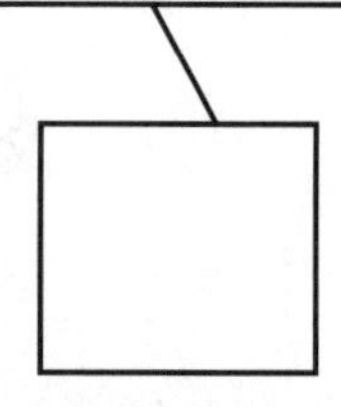

Common Trait: ____________

Reflect On The Process

2. How did you use classifying to help you make choices about how animals can be grouped together?

____

____

Name ______________________________ Date ______________

ANIMAL SKELETONS

Look at the skeletons. Can you tell what kind of animal each one belongs to? Try to label each one.

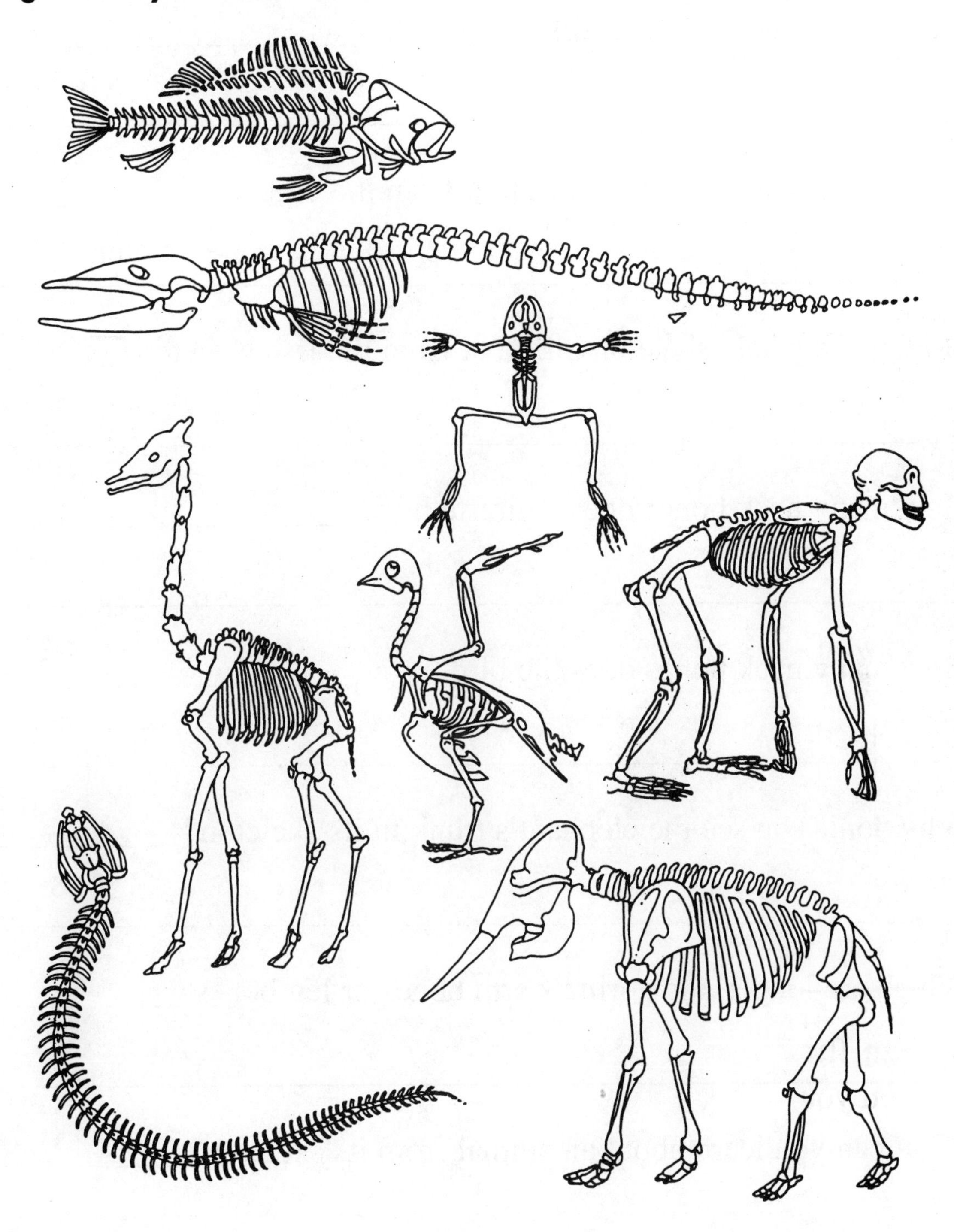

Go on to the next page.

Name ______________________ Date ____________

ANIMAL SKELETONS, P. 2

Use the skeletons to answer these questions.

1. How are all the skeletons alike? ______________________

____

2. How is the snake skeleton different from the others? ______________________

____

3. How is the whale skeleton different from the fish skeleton? ______________________

____

4. How many neck bones does a giraffe have? ______________________

____

5. How many neck bones does the bird have? ______________________

____

6. Why don't you see the elephant's trunk in its skeleton? ______________________

____

7. Which are longer, the gorilla's arm bones or leg bones? ______________________

____

8. What can you learn about an animal from its skeleton? ______________________

____

Name ______________________ Date ______________

WHAT DOES IT EAT?

Look at the pictures. Match each animal in Column 1 with its food in Column 2. Draw a line to connect them.

Column 1	Column 2

	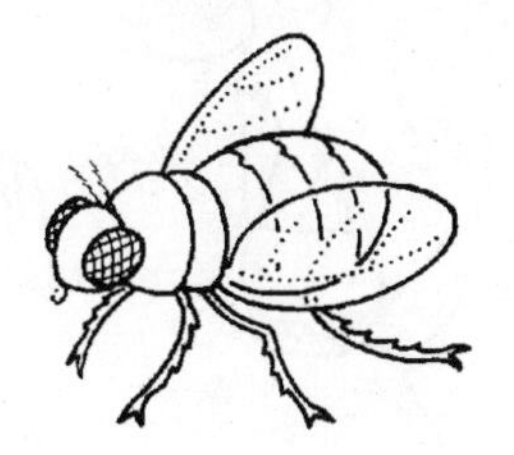

Name ______________________________ Date ______________

TEETH

Read this story about teeth. Then complete the activity below.

Use a mirror to look at your teeth. They are well-fitted for the food you eat. The front teeth, called *incisors,* are very sharp. They cut food. At the corners of your mouth are sharp, pointed teeth called *cuspids.* They tear food apart. Behind the cuspids are the *premolars.* These teeth tear and grind food. At the back of your mouth are large *molars.* Their flat tops make them good for grinding and crushing food. These four kinds of teeth help you chew foods from both plants and animals.

An animal's teeth must be fitted for the kinds of foods it eats. Meat-eating animals, such as tigers, have very large incisor and cuspid teeth. They use these teeth for catching and tearing their prey.

Plant-eating animals, such as cows, have sharp incisors. They use these teeth to cut plants. Their large molars grind the food into a pulp.

Look at the teeth of the animals below. If the teeth are good for eating meat, write *M* on the animal. If the teeth are good for eating plants, write *P* on the animal.

Name ______________________ Date ______________

FIT TO FIND FOOD

All animals are fit to find food in their environment. Different animals are fitted in different ways. Look at the pictures. Read about each animal. Decide which food each animal is fitted to eat. Write the name of the animal under its food.

1. A giraffe's long neck helps it to reach food high off the ground.

2. A butterfly's mouth forms a tube. It is used like a straw to suck liquid from a flower.

3. A starfish has hundreds of tiny feet. At the end of each foot is a sucker. The starfish sticks its suckers onto the shells of other animals. The feet then pull the shells apart.

Go on to the next page.

Name ______________________ Date ______________

Fit to Find Food, p. 2

4. A snake's jaw bones are held together loosely. When it is ready to eat, the snake makes these bones come apart. It can swallow an animal two or three times bigger than its head.
5. A land snail has a horny tongue covered with many hard teeth. It uses these teeth to scrape off bits of plants.
6. A mountain lion has sharp teeth and powerful jaws. By grasping, cutting, and tearing with its teeth, it can eat large animals.
7. A chameleon has a long, sticky tongue. It shoots out its tongue to catch insects.

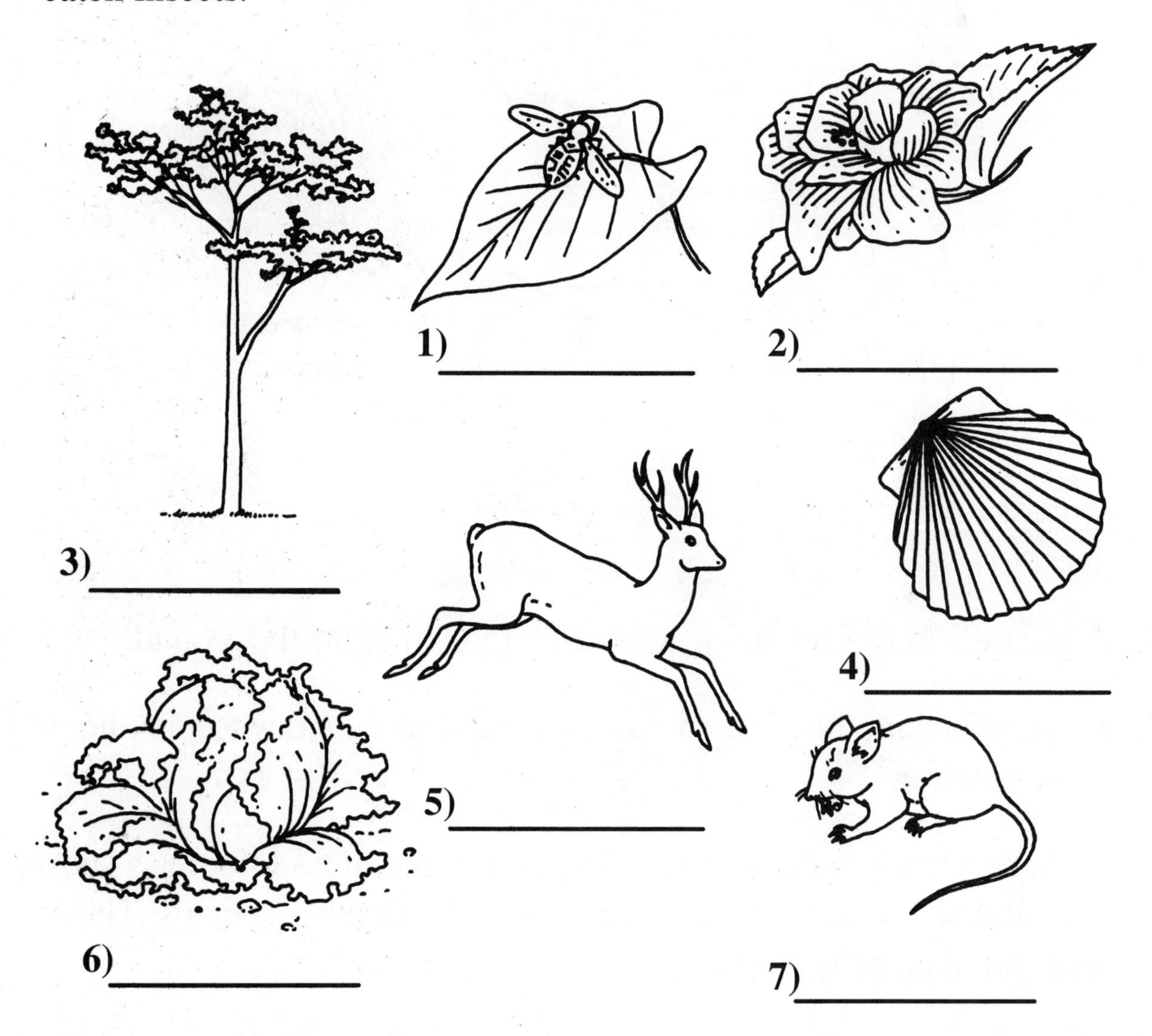

Name ______________________________ Date ______________

WHY ARE ANIMALS DIFFERENT?

Scientists know that beaks of birds are adapted for the different food each kind of bird eats. They also know that the beaks of birds are like tools. The heron uses its long beak like a pair of tweezers to snatch fish from the water. The sparrow uses its tough, blunt beak like a nutcracker to crack seeds. The woodpecker uses its strong, thick beak like a pickax to dig insects out of wood.

The beaks of different kinds of birds are adapted for eating different kinds of food. Think of a hawk's sharp, hooked beak. It is adapted for tearing.

Now design a bird's beak for eating one kind of food. You can wear your beak when you finish.

You will need:

sheet of white paper
colored pens and markers
white construction paper
transparent tape
scissors
string

1. Decide what kind of food you want the bird to eat. Then design a beak for eating that food.

2. Draw a picture of your bird and its beak on the white paper. Write the kind of food it eats.

Go on to the next page.

Name ________________________________ Date ______________

WHY ARE ANIMALS DIFFERENT?, P. 2

3. Fold the construction paper in half. Along the folded edge of the paper, draw the outside edge of your beak.
4. **CAUTION: Be careful using scissors.** Cut the folded paper along the line you drew. You now have both sides of your beak.
5. Tape the two parts of the beak together.
6. Color your beak, or add any design you like.
7. Tape a string to the ends of both sides of the beak.
8. Now put the beak on your face and tie the strings behind your head.

Answer these questions.

1. How is your bird's beak adapted for the food the bird eats? ________

____

2. You use your mouth and teeth to eat all kinds of food. How is a bird's beak different? ________________________________

____

3. The drawings show the beaks of five kinds of birds. Each beak has a different shape for a specific purpose. Under each beak is the name of the bird and the purpose of the beak. On the line under each bird, write the kind of food it eats.

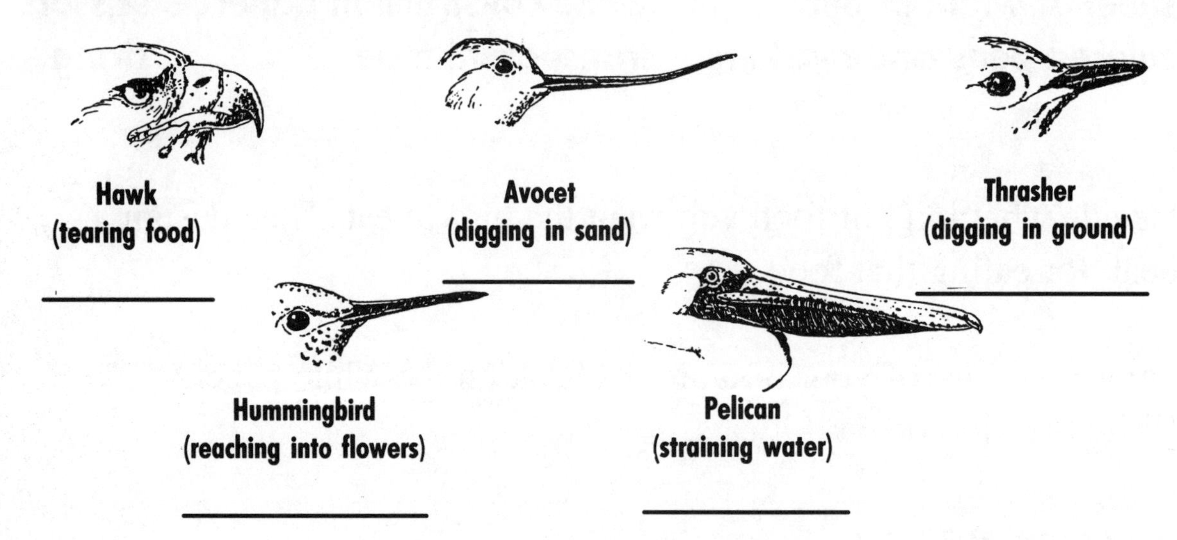

Name ______________________________ Date ______________

How Animals Behave

When young geese are born, they begin immediately to behave in certain ways. They don't have to learn these behaviors. They run around and peck soon after they hatch. Also, they follow their mother wherever she goes. Scientists call these behaviors *instincts.*

Dr. Konrad Lorenz of Austria studied the behavior of geese in their natural environment and in his home. He made a remarkable discovery about the behavior of young geese. He hatched some goose eggs. When the young geese were born, they had never seen their mother. The first thing they saw was Dr. Lorenz walking along. He was making clucking sounds. The young geese followed Dr. Lorenz. They thought he was their mother. Even when the young geese saw their mother, they paid no attention to her.

Dr. Lorenz found that instincts are important. In geese, the instinct to follow a mother is very strong. However, this instinct does not include a picture of what a mother looks like. This picture has to come from experience. A newborn goose will follow the first object that moves slowly and makes sounds like a goose.

The process in which newborn animals attach an instinctive behavior to something they see, such as their instinct to follow their mother, is called *imprinting*. Geese usually imprint on their mother. However, newborn geese will imprint on a wooden model of a goose, on other objects, or even on people.

Answer these questions.

1. Who was Dr. Konrad Lorenz?

____

____

____

Go on to the next page.

Name ______________________ Date ____________

How Animals Behave, p. 2

2. Where did Dr. Lorenz study geese?

____

____

3. What did he find out about instincts and animal behavior?

____

____

4. What instinct did he study in young geese?

____

____

5. What is imprinting? Why is it important?

____

____

____

6. Why did the newborn geese think Dr. Lorenz was their mother?

____

____

Do This

1. Read the book *Are You My Mother?* by P.D. Eastman (Random House, 1986). Go through the book slowly.

2. Make a list of all the different animal behaviors you can find.

Name ______________________________ Date ______________

The Best Defense

Read the paragraph, and then answer the questions that follow.

Animals exhibit a variety of behavior adaptations, both defensive and social, that help them survive. Animals behave in many different ways for many different reasons. Some animal behaviors help animals protect themselves from predators. Different animals have different ways of defending themselves. Some animals use camouflage to hide. Some, such as impalas, run away from predators. Some animals, such as giraffes, zebras, and musk oxen, stay together in large groups called herds. Staying in herds makes it harder for a predator to hunt just one animal. Some animals, such as porcupines and skunks, have body parts that help defend them. Opossums have a unique defense—they pretend to be dead.

Underline the best answer.

1. The way an animal behaves is
a. hibernation. **b.** migration.
c. an adaptation. **d.** protection.

2. For many animals, running away is their best
a. defense. **b.** behavior.
c. predator. **d.** camouflage.

3. Some animals keep safe by
a. eating only grass. **b.** living in groups.
c. swimming upstream. **d.** staying out in the open.

4. An opossum stays safe by
a. spraying strong-smelling liquids. **b.** pretending to be dead.
c. sticking out its quills. **d.** swelling up its body.

Name ____________________ Date __________

WHY DO THEY DO THAT?

- Animals have many different ways to defend themselves. Some of these include running away, hiding, and using body parts for defense.
- Some animals migrate, or move from place to place, in order to escape hot or cold weather, to find food, or to produce young.
- Some animals hibernate. Hibernation helps these animals survive very cold winters.

Complete the concept map about animal behavior. Fill in the blank spaces in the diagram using words from the word list.

WORD LIST

migration hibernation running away
cold weather predators

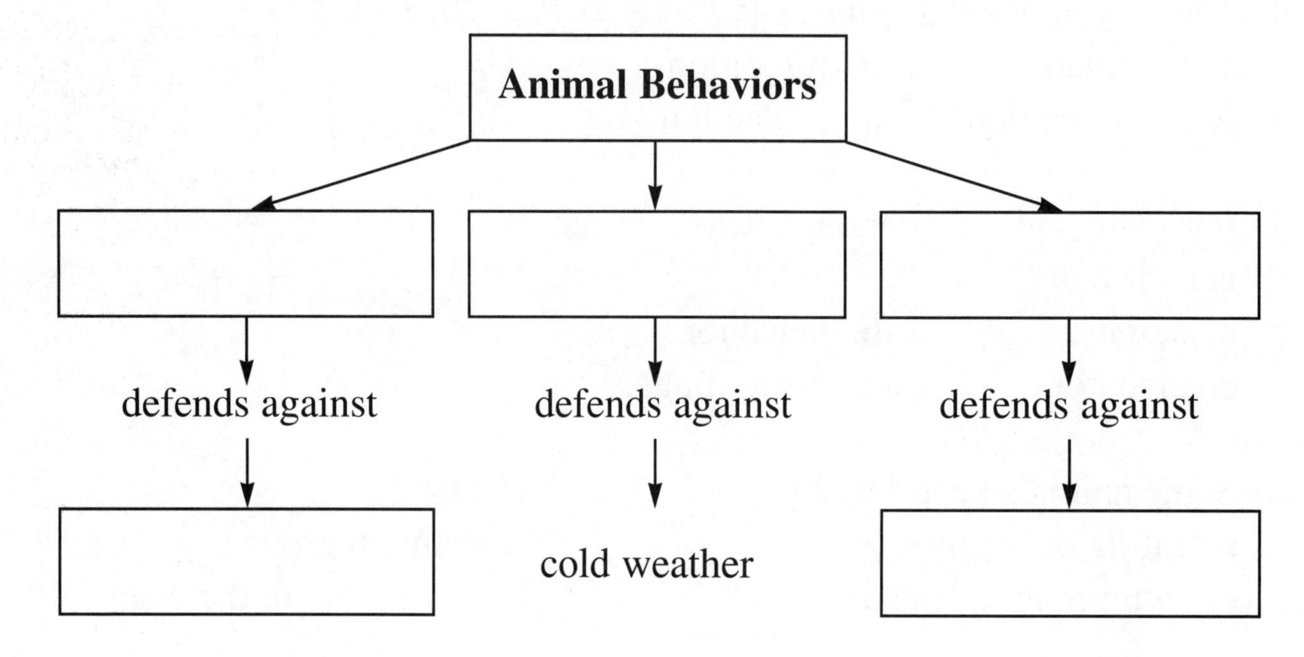

Name ______________________________ Date ______________

Animals Grow and Develop

Have you ever seen a koala at a zoo? Maybe you saw a mother koala with her young peeking out of her pouch. The koala is a *marsupial.* A female marsupial is a mammal with a pouch on her belly. Marsupials develop inside the mother's body for only a short time. At birth, a koala is poorly developed and almost helpless. It has no fur, and its skin is pink. It cannot see. It is only about as long as the tip of your finger and weighs less than the eraser on a pencil. As soon as it is born, the tiny koala crawls several centimeters up through its mother's fur into her pouch.

Only one koala is born at a time. Once inside the pouch, the koala fastens its mouth onto a nipple to get the milk it needs. The young koala stays attached to the nipple inside the pouch for five to six months. Then it climbs out of the pouch, but it still clings to the mother's chest. Next the mother begins to feed her young some eucalyptus leaves. In a few weeks, the young koala stops drinking milk and eats only leaves.

Why do most people have to go to the zoo to see a koala? Koalas and many other kinds of marsupials live only in Australia and on the islands around it. Only one marsupial, the opossum, lives somewhere else. Opossums live in North and South America.

If you have seen a newborn kitten, you know it is very different from a newborn koala. Usually, four or five kittens are born at one time. They leave their mother's body at birth. They cannot see or hear. So they depend on their mother to nurse, clean, and protect them. Each kitten finds one of its mother's nipples to drink milk it needs. The kittens do not stay attached to the nipple, though. After 7 to 10 days, the kittens' eyes open. Their teeth begin to appear. At about three weeks, they start to walk and explore. About a week later, they begin to eat solid food.

Go on to the next page.

Name ______________________________ Date ______________

Animals Grow and Develop, p. 2

Answer these questions.

1. Why must a koala crawl into its mother's pouch and stay there after it is born?

____

____

____

2. When is a young koala able to leave its mother's pouch?

____

3. How does the food of the young koala change when it leaves the pouch?

____

4. How is a koala at birth different from a kitten at birth?

____

____

5. How are the first four weeks of life different for a kitten than for a young koala?

____

____

____

You will need:

map of the world   construction paper   pen with a heavy tip

Do This

1. Use classroom books or books from the library to find where marsupials live. Then, on your map, find these places.

2. On construction paper, draw these places. Write the names of the places you draw. In each place, write the name of the marsupial that lives there. Hang your finished work in your classroom.

Name ______________________________ Date ______________

Animal Babies

Animals have many different ways of producing young. Some animals give birth to live young. Some animals lay eggs. Eggs provide a wet environment in which the young of some animals can grow. Animals that care for their young have fewer babies than animals that do not care for their young. The young of some animals, like cats, look very much like their parents. The young of other animals, such as butterflies, do not look like their parents.

Use words in the word list to fill in the steps in the life cycle of a butterfly.

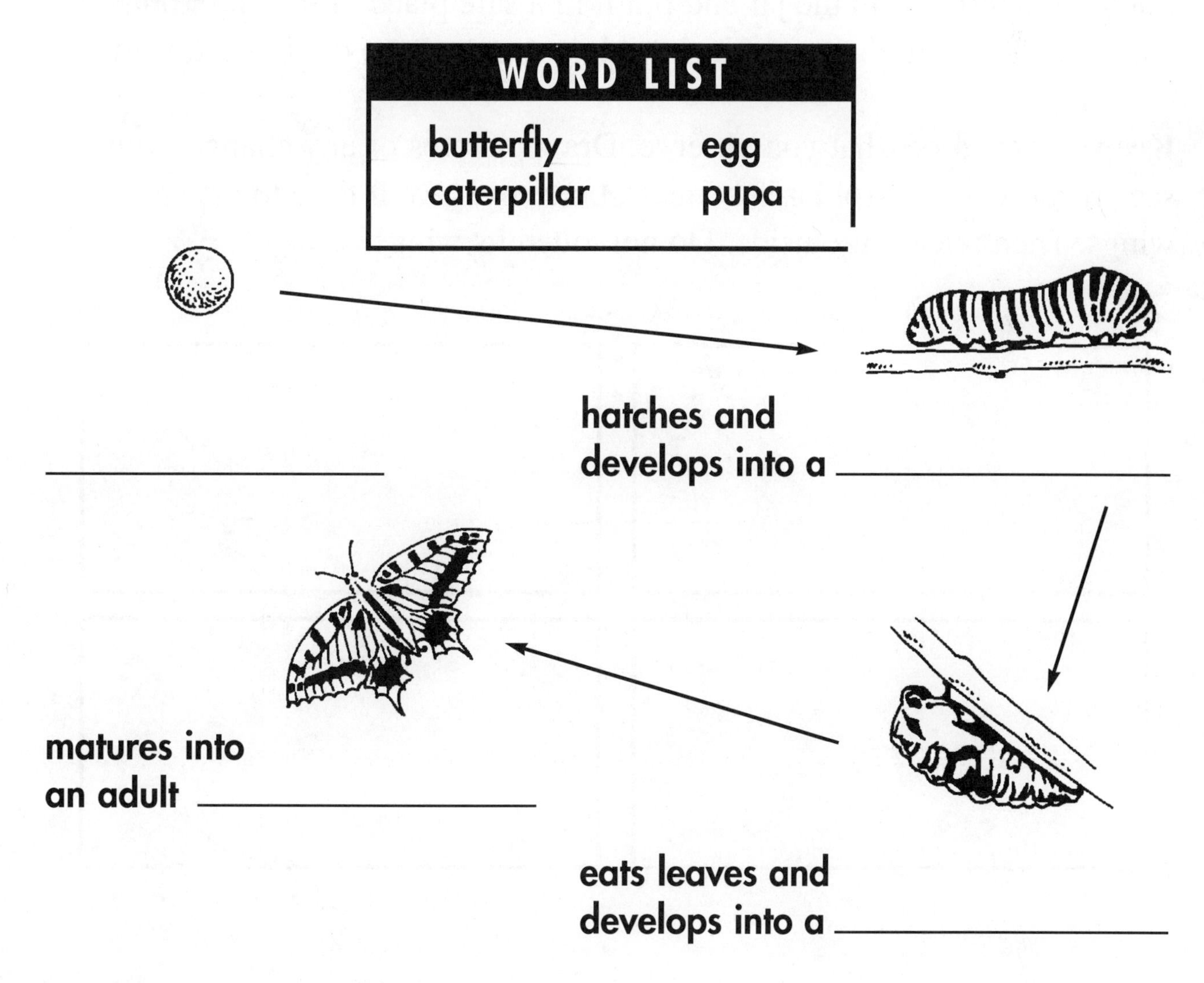

Go on to the next page.

Name ______________________________ Date ________________

ANIMAL BABIES, P. 2

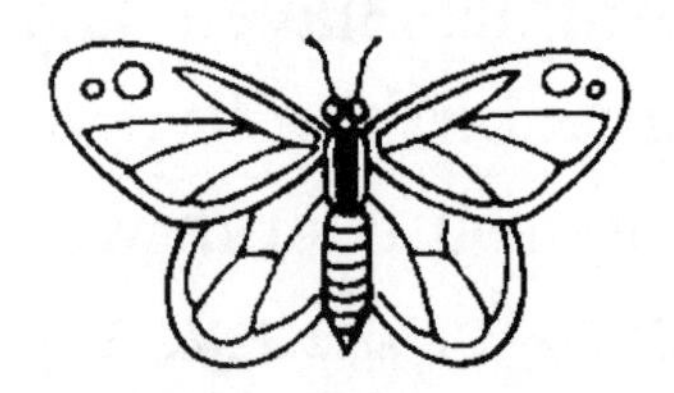

You can watch a caterpillar turn into a butterfly in your classroom.

You will need:

glass jar with lid; small holes poked in lid for oxygen supply
small leafy branch that will fit into jar
caterpillar

Put your caterpillar in the jar and put it in a safe place in the classroom where nothing will disturb it. It should not become too cold or too warm.

Keep a record of what you observe. Draw pictures of any changes you see. When your caterpillar becomes a butterfly, give it time to dry its wings. Then release it outside. Do not touch its wings.

Name ______________________________ Date ________________

ANIMAL ADAPTATIONS

In each box below is a word that describes an adaptation that helps the animals in the pictures live in streams. Draw a line from each animal to the adaptation that it uses. Use a different colored pencil for each animal.

beak	body shape	color	contour feathers
down feathers	feet	fins	long legs
scales	spots	tongue	wings

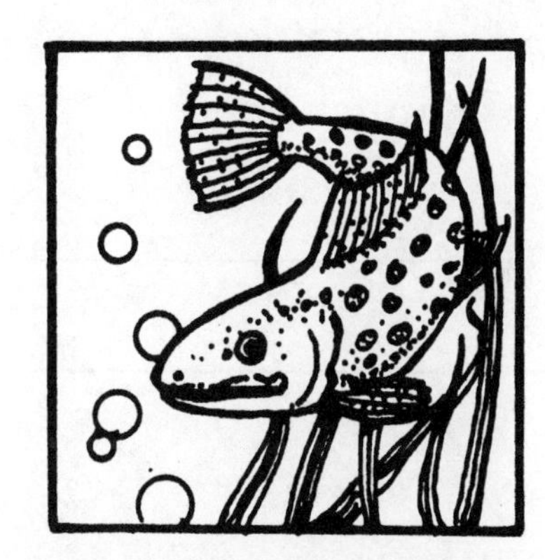

Go on to the next page.

Name ______________________________ Date ______________

ANIMAL ADAPTATIONS, P. 2

For each adaptation listed below, write a sentence on the lines, explaining how one of the animals in the drawings uses that adaptation to survive.

1. long legs

____

____

2. spots

____

____

3. tongue

____

____

4. wings

____

____

5. fins

____

____

6. beak

____

____

Name ______________________ Date ____________

Animal Dot-to-Dot

A. Follow the dots to make pictures of animals.

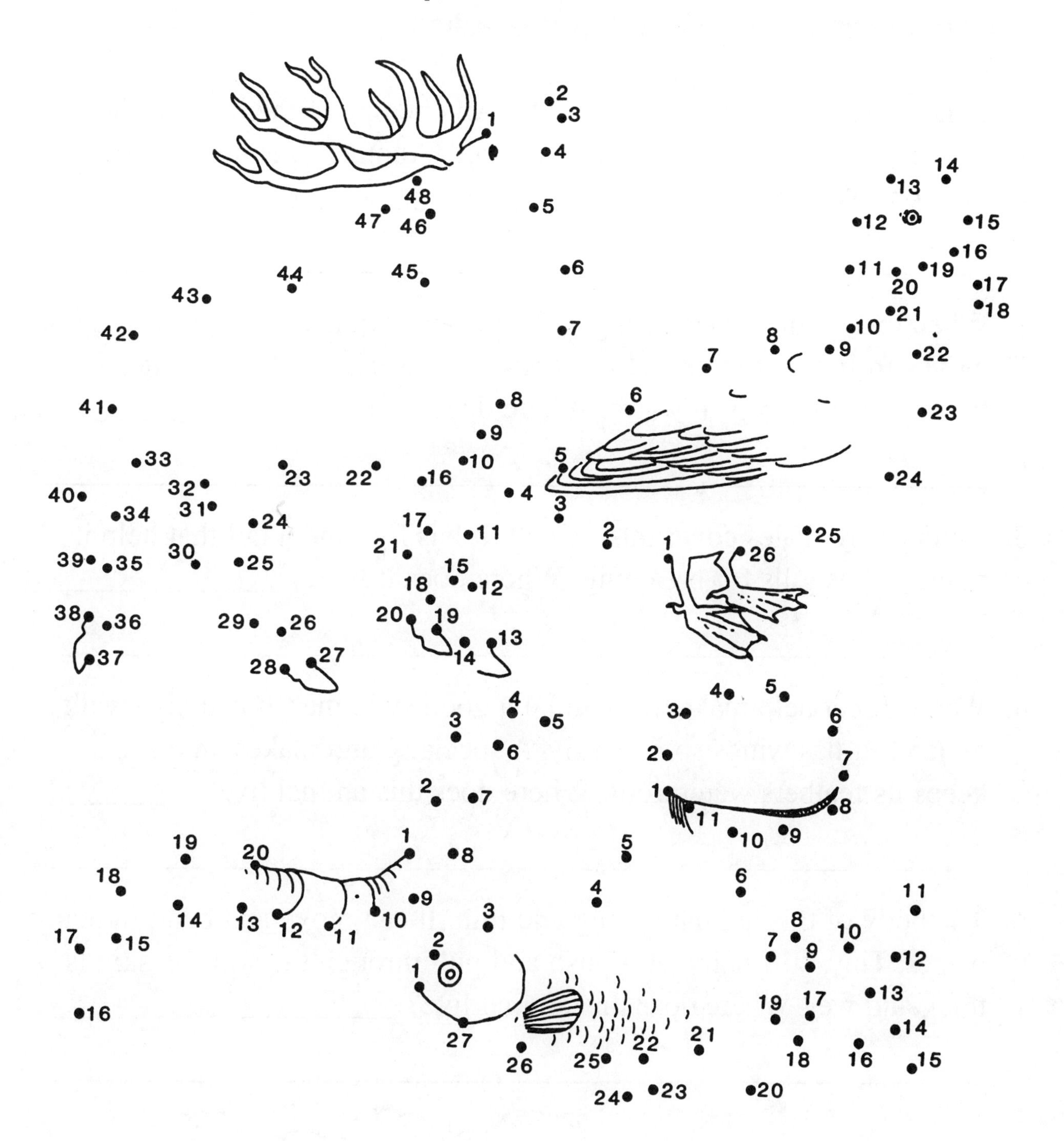

Go on to the next page.

Name ______________________________ Date ______________

ANIMAL DOT-TO-DOT, P. 2

B. Read about the animals. Each one is fit to live in its own environment. Then answer each question.

1. This animal lives inside a hard shell. An opening in the back of the shell lets in water filled with bits of food. Another opening lets water flow out. Where does it live? ______________________________

____

2. When this animal smells danger, it runs away quickly. When feeding, it moves from tree to tree, eating twigs, berries, and buds. In winter, its thick short fur keeps it warm. Where does it live? ______________

____

3. Hard, slimy scales cover this animal. It has fins and a tail that help it move. It has gills for breathing. Where does it live? ____________

____

4. Webbed feet help make this animal a good swimmer. It can also walk on land. It has wings so it can fly. A special gland makes an oil that keeps its feathers waterproof. Where does this animal live? ________

____

5. The body of this animal is long and thin. It can move and bend in many places. Tiny, stiff hairs let it push and pull through the soil. Its skin is thick and wet. Where does this animal live? ____________________

____

Name ____________________ Date ____________

Where Does It Live?

Look at the map. It shows you where different environments are located. Then look at the pictures of the animals.

Decide in which environment each of these animals lives. The pictures will give you hints. Write the number of each animal on the part of the map where it belongs.

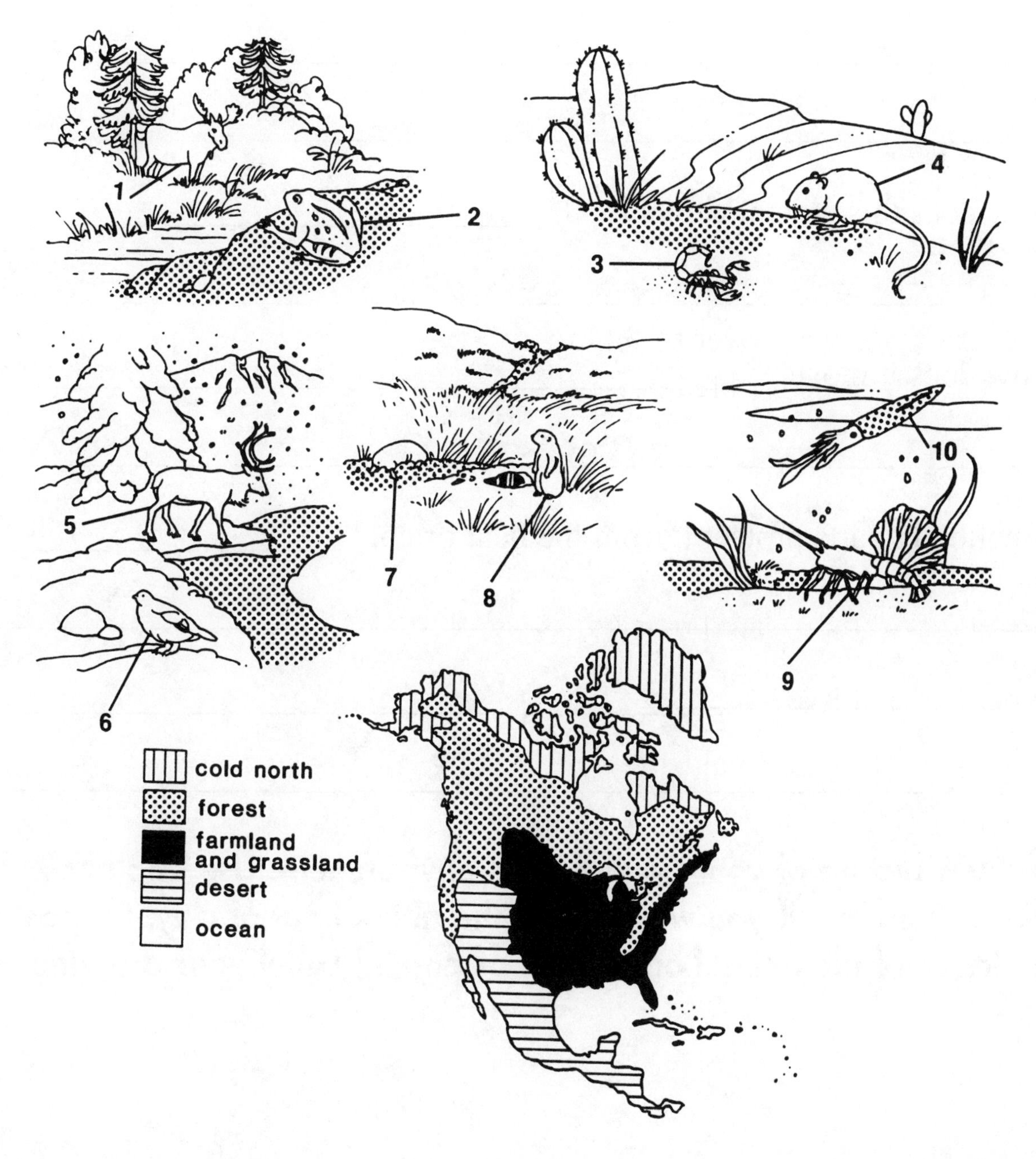

Name ______________________________ Date ______________

ANIMAL FACTS

A. Pick an animal you want to study. Learn how it is fitted to live in its environment. To find out about your animal, look in your science book or a library book. Then answer these questions.

1. What body parts help to keep this animal safe from enemies?

____

____

2. How does this animal get its food? ______________________

____

3. How does it move? ______________________________

____

4. What body parts protect it from the heat or cold? ____________

____

5. Where does it live? ______________________________

____

B. Draw a picture of your animal in its environment. Use another piece of paper. (If you wish, make the animal out of clay. Or paste a picture of the animal on a sheet of paper.) Label your drawing.

Name ________________________________ Date ________________

HIDING OUT

These animals live in different places. The kangaroo rat lives in the sandy desert. The green snake lives in the grassland. The white rabbit makes its home in the snow-covered Arctic. The *color* of each animal helps it to be safe in its environment. **To see why, you will need:**

green crayon **brown crayon**

1. Color the land in the desert brown. Color the grass in the grassland green. Leave the snow in the Arctic white.

2. Color the kangaroo rat with the same crayon you used to color the desert. Color the snake green. Leave the rabbit white.

Go on to the next page.

Name ______________________________ Date ______________

HIDING OUT, P. 2

Answer these questions.

1. How does the color of the kangaroo rat keep it safe in the desert?

____

____

2. How does the color of the snake keep it safe in its home? __________

____

____

3. How does the color of the rabbit keep it safe in the Arctic? _________

____

____

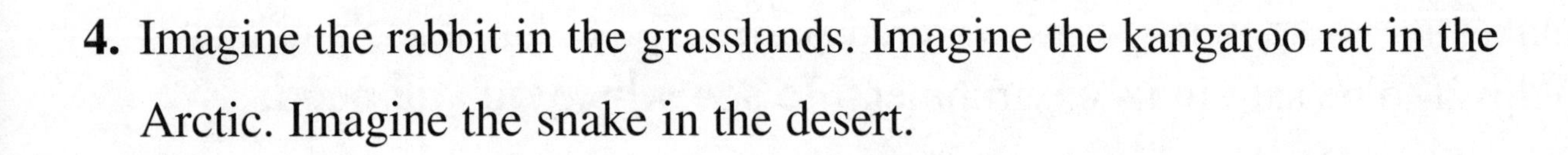

4. Imagine the rabbit in the grasslands. Imagine the kangaroo rat in the Arctic. Imagine the snake in the desert.

5. Would these animals be safe from enemies in these other environments?
________Why or why not? ______________________________________

____

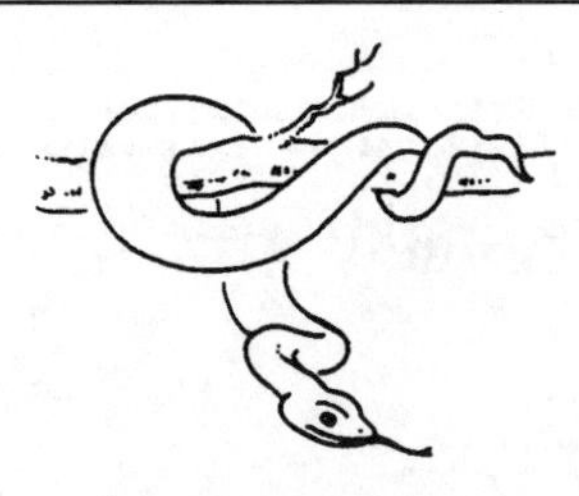

Name ______________________________ Date ______________

Fit to Live in the Cold

The Arctic is a very cold place. Ice and snow cover the land for most of the year. Yet snowy owls and polar bears make their homes there. These animals are fit to live in the Arctic. A bear has fur to help keep it warm. An owl has down feathers. Down feathers are soft and fluffy. They are close to the bird's body.

To see how the bear and the owl keep warm, you will need:

3 jars of the same size	**water**
thermometer	**clock**
down vest (or jacket)	**fur hat (or plush material)**

1. Fill each jar to the same level with very warm water. Measure the temperature of each. All three jars must be the same temperature. If they are not, pour out a little water from the cooler ones. Then add hot water and stir. Keep doing this until all the jars are the same temperature. Record the starting temperature in the chart.

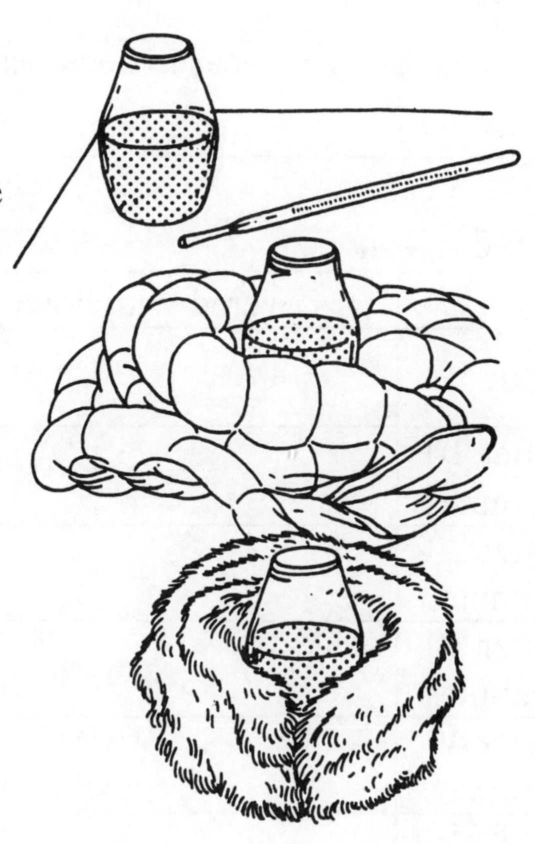

2. Wrap one jar in the down vest or jacket. Wrap another jar in the fur (or plush) hat. Leave the last jar uncovered.

3. Measure the temperature of each jar every 10 minutes for an hour. Record your results in the chart.

Go on to the next page.

Name ______________________ Date ____________

Fit to Live in the Cold, p. 2

Answer these questions.

1. Which jar lost the most heat? ______________________

2. Which stayed warmer, the fur-covered jar or the down-covered jar?

____

3. How does a polar bear's fur keep it warm? ______________________

____

4. How does a snowy owl's down feathers keep it warm? ____________

____

5. What can you wear to keep warm in cold temperatures? ____________

____

TIME	WATER TEMPERATURE		
	jar covered with down	jar covered with fur	uncovered jar
start			
after 10 minutes			
after 20 minutes			
after 30 minutes			
after 40 minutes			
after 50 minutes			
after 60 minutes			

Name ______________________ Date ______________

A Swampy Home

Read this story about alligators. Then answer the questions.

Alligators are reptiles. They live in rivers and swamps. The swamps are wet lands where water doesn't drain away. After a long time without rain, a swamp may begin to dry up. To stay wet all year, alligators build "gator holes." An alligator does this by pushing mud aside with its feet and body. It can dig a hole the size of a small swimming pool. Gator holes rarely dry up. Fish and frogs share the alligator's hole. If they are not careful, they may become the alligator's meal!

At one end of the gator hole, the alligator digs a large tunnel. This is where the alligator rests on cold winter days.

In the spring, the female alligator builds a nest on land. With her body and tail, she scrapes mud and plant material into a large mound. With her back legs, she digs a hole on the top of the mound. She lays her eggs in this hole. Then she covers them. The plant materials keep the eggs warm until they hatch.

Go on to the next page.

Name ______________________________ Date ______________

A Swampy Home, p. 2

Read each sentence. If it is true, write *T* in front of it. If it is false, write *F* in front of it.

1. _____ Alligators are reptiles.

2. _____ Alligators livc in rivers and swamps.

3. _____ A gator hole is a deep hole that stays dry all year.

4. _____ Fish and frogs often share an alligator's hole.

5. _____ An alligator rests in a tunnel on hot summer days.

6. _____ The female alligator builds a nest in the gator hole.

7. _____ The female covers the eggs.

8. _____ The plant material in the nest keeps the eggs warm until they hatch.

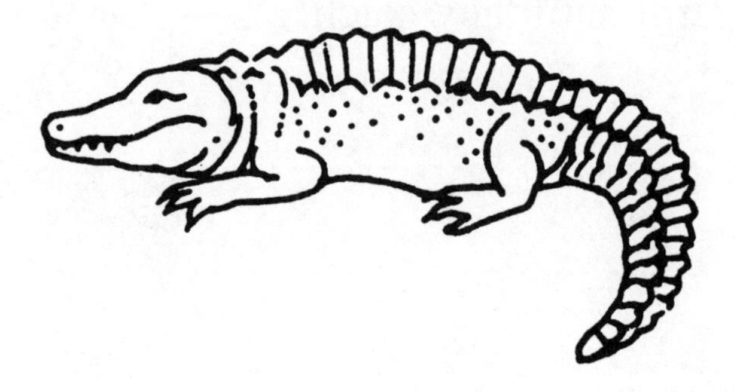

Name ______________________ Date ____________

NESTS ARE OUT OF SIGHT

Bird's nests can be found almost anywhere. They can be hidden in caves, behind rocks, or balanced on tree limbs. Most birds use their nests only to raise their young. A few, such as woodpeckers and owls, go to their nests to sleep and to escape their enemies.

A. Look at the pictures. Then, on the next page, read about the kind of nest each bird builds. Match the birds with their nests. Write the correct number in the space under each picture.

Go on to the next page.

Name ______________________ Date ______________

NESTS ARE OUT OF SIGHT, P. 2

1. The cliff swallow uses mud to build its nest on the side of a building. It lines the nest with grass and feathers.

2. A robin molds mud, grass, and twigs together to form its nest. It lines the nest with soft grass.

3. On a cliff, building ledge, or bridge, you may see a pile of sticks, paper, and other materials. This may be a pigeon's nest.

4. A hummingbird weaves its nest from soft parts of plants. It covers the nest with plants called lichens. Spider silk holds the nest to a tree limb.

B. How hard is it to build a bird's nest? To find out, get some things birds use to make their nests. You will need:

twigs	**leaves**	**grass**	**mud**
clay	**string**	**ribbon**	**strips of paper**

See if you can make a nest that stays together. Compare your nest with the ones made by your classmates.

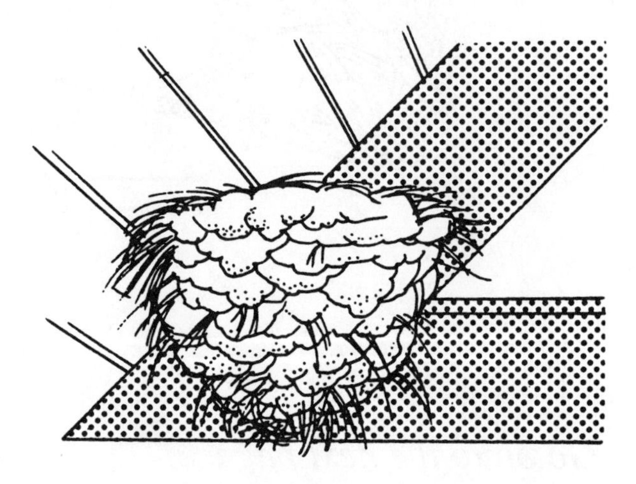

Name ______________________________ Date ______________

SUNFISH NESTS

Read this story about sunfish. Then answer the questions.

Along the edge of many rivers and lakes, tiny craters dot the sandy bottom. These craters are sunfish nests. The nests are safe places for sunfish eggs to hatch.

The male sunfish is the nest builder. First the male finds a safe place for the nest. Then he swims in circles, fanning the water with his fins and tail. The powerful strokes move the sand. A small hole forms. The nest is finished.

After the nest is made, the female arrives. She lays eggs in the nest and leaves. The male guards the nest.

1. Why do sunfish need a nest? ______________________________

____

2. Which fish builds the nest, the male or the female? ______________

3. What does the sunfish do before building the nest? ______________

____

4. How does the fish make the nest? ______________________

____

5. What does the female sunfish do? ______________________

____

6. Which fish guards the eggs? ______________________________

____

Name ______________________ Date ____________

A Spider's Web

Read about spiders and their webs. Then you will make a web.

Bzzzzz! Bzzzzz! A fly is trapped in the sticky web of a garden spider. The fly struggles to free itself. The web shakes. From its lookout at the web's center, the spider zooms in. It wraps the fly in silk. It bites the fly and poisons it. Then, if the spider is hungry, it eats the fly. If not, it saves its meal for later.

Spiders have small glands inside their bodies. Some glands make sticky silk threads. Others make dry threads. Inside the spider, the silk is liquid. The silk hardens as it leaves the body.

Without being taught how, a spider spins its web. First, it builds a frame of dry silk. These dry threads hold up the web. They also make a safe place for the spider to walk. Next, the spider lays down sticky threads. It weaves them into a close-knit trap.

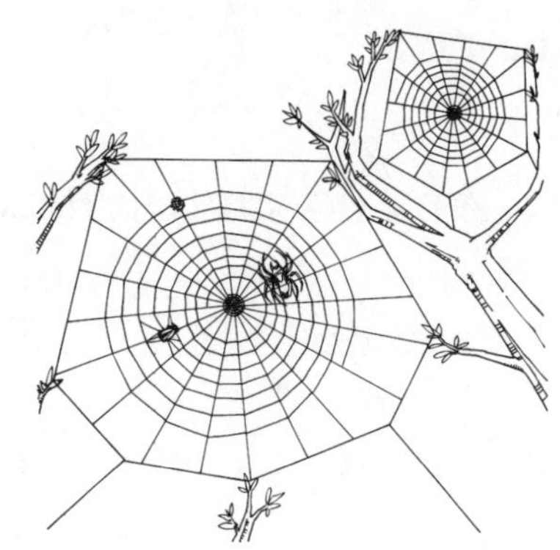

You can make a web like a spider's. You will need:

scissors	**glue**
string or yarn	**heavy paper**

1. Glue down string to form the outside of the web. Then add pieces of string from the outside toward the center. These pieces should form spokes like those of a bicycle wheel.

2. Glue pieces of string in tiny circles. Start near the center. Work your way around until the web is filled in. Compare your web with the web in the picture.

Name ______________________________ Date ______________

INFERRING

Inferring is using what you have observed to explain what has happened. An observation is something you see or experience. An inference is an explanation of an observation, and it may be right or wrong.

Think About Inferring

Ask yourself:
1. What observation am I trying to explain?
2. What information do I need so I can tell whether my explanation is correct?
3. How can I decide whether my explanation was correct?

Try It

Sam found a tiny green caterpillar outside. He brought it home and put it in a jar with leaves for it to eat. He sprinkled the leaves with water. Then Sam covered the jar, using a lid with holes punched in it. He put the jar on a sunny windowsill. Sam added more leaves and water whenever the caterpillar needed them. The caterpillar grew bigger and bigger.

1. Make an inference about what made the caterpillar grow.

____

____

Sam came home from a weekend trip. There was no caterpillar in the jar. Instead, a pale, grayish sac hung from the lid of the jar.

2. Infer what has happened to the caterpillar.

____

____

Go on to the next page.

Name ______________________ Date ______________

INFERRING, P. 2

Sam always looked in the jar when he woke up. One morning, he looked in the jar and saw that the sac was broken open. Inside the jar was a white and brown moth.

3. Infer where the moth came from.

____

____

4. What information did you use to make your inference?

____

____

5. How would you test the inference you made in situations 1, 2, and 3?

____

____

____

____

____

____

Reflect On The Process

6. How did you use inferring to help you understand what happened to Sam's caterpillar?

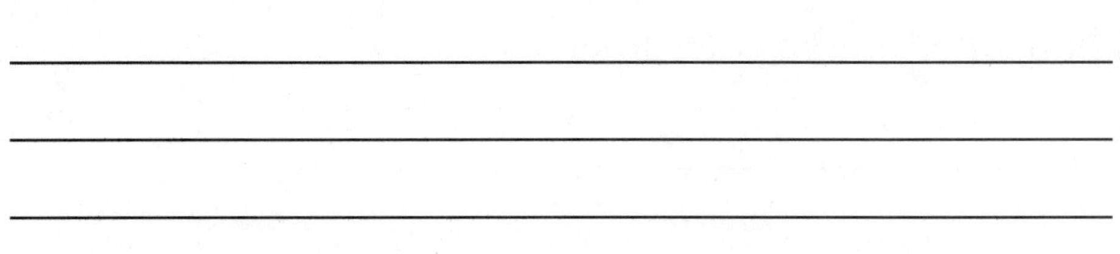

Name ________________________________ Date ______________

PREDICTING

A prediction is a statement about what you think will happen. To make a prediction, you think about what you've observed before. You also think about how to interpret the data you have.

Think About Predicting

Ask yourself:

1. What information am I using to decide what I think will happen?
2. How will I gather more information about what will really happen?
3. How close was my prediction to what actually happened?
4. If my prediction was not what really happened, what have I learned from it? What new predictions can I make?

Try It

Algae and other plants

A. This is a drawing of a food web in a stream. Suppose the stream becomes polluted with phosphates. Phosphates cause algae to grow faster.

Go on to the next page.

Name ______________________________ Date ______________

PREDICTING, P. 2

1. Predict what might happen to the algae. Predict what might happen to the perch. ______________________________

2. What information did you use to make your prediction?

3. What kind of test can you do to find out what will really happen?

B. Suppose a disease kills all the herons.

4. Predict what will happen to the perch and frog population. What might happen to the insect population? ______________________________

5. What information made you think that?

Reflect On The Process

6. How did you use predicting to help you figure out what would happen in each situation? ______________________________

Name ______________________________ Date ______________

Animal Crossword

Use the words in the box to complete the sentences. Then fill in the puzzle.

amphibian	backbone	birds	classes	earthworm	environment
fish	gills	insects	lobster	mammals	mollusks
nerve cord	reproduce	reptile	scales	skeleton	starfish

ACROSS

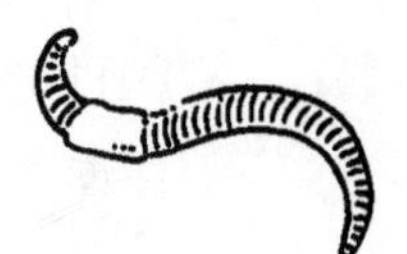

2. The biggest class of animals with jointed legs is the ____________ class.
3. An animal with spiny skin is a ______________________.
6. An animal's surroundings is called its ____________________.
8. ____________________________ allow fish to breathe underwater.
9. A ______________________ has jointed legs and lives in the ocean.
13. Bones that support an animal are called its _________________.
14. Animals that give birth to live young are ___________________.
15. Animals that are alike in some ways are grouped in ____________.
16. Animals that have feathers are ______________________.
17. A chain of small bones that can bend is the _________________.

Go on to the next page.

Name ______________________________ Date ______________

ANIMAL CROSSWORD, P. 2

DOWN

1. Inside the backbone is the ______________________________.
4. Frogs and salamanders belong to the ____________________ class.
5. ____________________ use their tails and fins to steer through water.
6. An ______________________ is a ringed worm that lives on land.
7. Most __________________ live in the ocean and have hard shells.
10. A turtle belongs to the ______________________________ class.
11. Animals can __________________ to make more of their own kind.
12. ______________________________ cover fish and some reptiles.

Name ______________________________ Date ______________

Unit 3 Science Fair Ideas

A science fair project can help you to understand the world around you better. Choose a topic that interests you. Then use the scientific method to develop your project. Here is an example:

1. **PROBLEM:** Why do different birds have such different beaks?
2. **HYPOTHESIS:** Models of different types of beaks and the foods that different birds eat can help show why the beaks are different.
3. **EXPERIMENTATION:** Look up several different types of birds. Observe the different types of beaks and note what each bird eats and where it gets its food. Observe birds in nature. Make models of the different beaks out of clay, wood, papier mache, or any material that you think will work well. Label the beaks and tell what foods the birds eat and how they get food.
4. **OBSERVATION:** The materials show the different types of beaks and the foods that the birds eat.
5. **CONCLUSION:** Models of birds' beaks are a good way to show the differences between them and how they are adapted to get the foods each bird needs.
6. **COMPARISON:** Conclusion agrees with hypothesis.
7. **PRESENTATION:** Display the models with your project. Label and describe each type of beak and describe the way it works to help each bird.
8. **RESOURCES:** Tell of any reading you did to help you with your experiment. Tell who helped you to get materials or set up your experiment.

Other Project Ideas

1. How are animals adapted to their environments?
2. Could an animal from one environment live in a different environment?
3. What happens when an animal is removed from the food chain?
4. How do different animals care for their young?

UNIT 4: HEALTH
BACKGROUND INFORMATION

Healthy Bodies

Health for children revolves around healthy foods, plenty of exercise, and good hygiene. As children grow, they should begin to recognize that they can make choices that will help them live healthy lives. They need to learn the connections between what they eat and the way they look and feel. They need to have the basic information that will help them to make good food choices. Children need to know that it is never too early to begin healthy habits in eating, exercise, and hygiene. The habits they form now will affect their lives for many years to come.

Nutrition

The body needs to receive certain nutrients in order to grow and to stay healthy. These nutrients are broken down into six types: carbohydrates, protein, fat, vitamins, minerals, and water.

- Carbohydrates are sugars and starches. Sugars, such as fruits and honey, give the body quick energy while the starches, such as bread, cereal, and rice, give the body stored energy.
- Proteins come from foods such as milk, cheese, lean meat, fish, peas, and beans. They help the body to repair itself. Proteins are used by the body to build muscle and bone, and they give the body energy.
- Fat is important for energy, too, and it helps to keep the body warm, but if the body does not use the fats put into it, it will store the fat. Fats come from foods such as meat, milk, butter, oil, and nuts.
- Vitamins are important to the body in many ways. Vitamins help the other nutrients in a person's body work together. Lack of certain vitamins can cause serious illnesses. Vitamin A, for example, which can be gotten from foods such as broccoli, carrots, radishes, and liver, helps with eyesight. Vitamin B from green leafy vegetables, eggs, and milk, helps with growth and energy. Vitamin C from citrus fruits, cauliflower, strawberries, tomatoes, peppers, and broccoli, prevents sickness.
- Milk, vegetables, liver, seafood, and raisins are some of the foods that provide the minerals necessary for growth. Calcium is a mineral that helps with strong bones, and iron is needed for healthy red blood.
- Water makes up most of the human body and helps to keep our temperature normal. It is healthy and recommended to drink several glasses of water each day.

Foods have long been divided into four basic food groups-meat, milk, vegetable-fruit, and bread-cereal. New discoveries have led to a change in the divisions so that in a food pyramid, fruits and vegetables are separated, and fats are included at the top of the pyramid. The recommended servings for each group have also changed over time. Eating the right amount of foods from each group each day gives one a balanced diet. Eating too many foods from one group or not enough of another can lead to deficiencies or weight problems. Although vitamin supplements can help with these deficiencies, vitamins are best absorbed in the body naturally through the digestion of the foods that contain them.

- The Bread-Cereal (Grain) Group contains foods made from grains such as wheat, corn, rice, oats, and barley. Six to eleven servings from this group each day give you carbohydrates, vitamins, and minerals.

- The Vegetable and Fruit Groups contain vitamins, minerals, and carbohydrates. Two to four servings of fruits and three to five servings of vegetables each day are recommended.

- The Meat Group includes chicken, fish, red meats, peas, nuts, and eggs. The meat group contains much of the protein we get from our diets, but it also includes fats. Two to three servings from the meat group each day are recommended.

- The Milk Group includes milk (whole and skim), butter, cheese, yogurt, and ice cream and gives us fat, vitamins, protein and minerals that are important for strong bones and teeth, such as vitamin D. Two to three servings from the milk group each day are recommended.

- The Fats, Oils, and Sweets Group, including butter, oil, and margarine, should be used sparingly.

Hygiene

Keeping the body clean is an important part of staying healthy. Children need to know that when they wash, they are washing off viruses and bacteria, or germs, which can cause illness. Washing the hair and body regularly prevents bacteria from entering the skin through cuts and from getting into the mouth. Hands should always be washed after handling garbage or using the bathroom.

Germs can also come from other people. Children should be discouraged from sharing straws, cups, or other utensils. They should be reminded to always cover their mouths when they sneeze or cough, and to use tissues frequently. Children also need to be reminded not to share combs or hats.

Teeth

Regular brushing and flossing can help keep teeth healthy. Avoiding sweets will also help. Most children have all their baby teeth by the time they are two years old. When they are about seven, they will begin to lose their baby teeth and permanent teeth will begin to appear. Although the baby teeth will fall out, it is important to take good care of them and the gums that surround them.

The part of the tooth that you can see is called the crown. The neck of the tooth is between the crown and the roots. The gums hold teeth in the mouth. Teeth are made up of three layers, the enamel, the dentin, and the pulp. The enamel is the white surface that we see. The dentin is a bone-like substance under the enamel. The pulp inside the tooth contains the blood vessels and nerves.

Decay is caused by acids in the mouth that eat into the enamel. The acids are caused by bacteria that live on the food in your mouth. If you brush and floss regularly, the food is taken out of your mouth, and the bacteria cannot live there. When you brush, you remove the plaque from your teeth, as well. Plaque is the sticky yellow film that develops on your teeth from food, bacteria, and acid. Decay can cause a hole in the tooth called a cavity. It can also harm the gums and cause gum disease. Regular dental exams and x-rays will detect any decay that you may have missed.

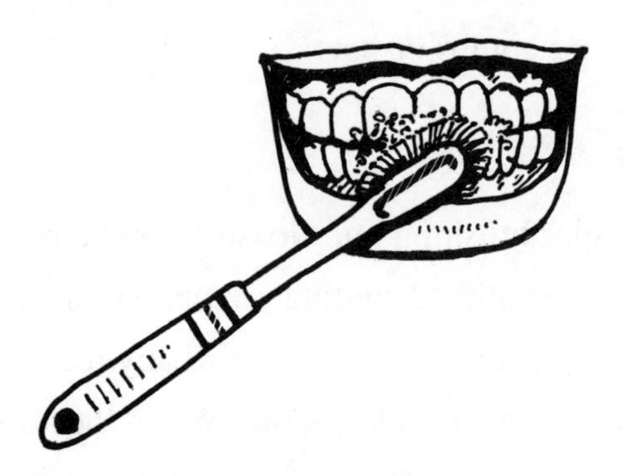

Exercise and Sleep

There are over 600 muscles in the human body. Muscles are bundles of tissue that respond to nerve impulses by expanding and contracting. When they expand and contract, they make the blood circulate, move food, expand and contract the chest for breathing, and move the outer parts of the body.

Muscles grow when they are used, and contract when they are not used. Muscles that become unaccustomed to exercise can be injured by sudden or strenuous activity. This is why muscles should be exercised regularly and in moderation. Occasional strenuous activity is not advantageous to the muscles and does not give long-term results. Exercising the muscles makes the body grow larger and stronger and helps make the heart strong.

Regular exercise can relax the body and help people get a good night's rest. Sleep is an important part of keeping the body healthy. People need different amounts of sleep at different times of their lives. Babies sleep most of the time because their bodies are growing very quickly. School children usually require from eight to ten hours of sleep, and adults need about seven or eight hours. Sleep allows the body and mind to rest. If we don't get enough sleep, our bodies and minds do not function as well as they should. Our attention wanders, and we become forgetful. Our muscles will feel weak and less coordinated. Lack of sleep can also make people irritable and impair their judgment.

Health and the Environment

Today, more than ever, people must be aware of the effect of the environment on their health. Pollutants in the air and water, excessive exposure to the sun or cold, a high pollen count in the air, and breathing in second-hand smoke are examples of environmental hazards. Some of these are more easily within our control, such as being sure to wear sunscreen when outdoors and avoiding prolonged exposure to the sun. The pollutants around us may be more difficult to avoid, but students should be made aware of their existence and the importance of each person doing his or her part in working to minimize such hazards. Students can have an effect on their own environment when they pick up trash and show respect for their surroundings. Our air and water are necessary to all life on Earth, and students should recognize the importance of keeping them clean.

The environment also presents some natural problems, such as insects that bite and sting, and poisonous plants. Besides causing you pain and irritation, mosquitoes can carry organisms that cause diseases from one person to the next. Ticks can cause lime disease. Poisonous plants such as poison ivy, poison oak, and poison sumac should also be avoided. Students should look them up and be able to recognize these plants. All three can cause painful, itchy rashes that can last for weeks.

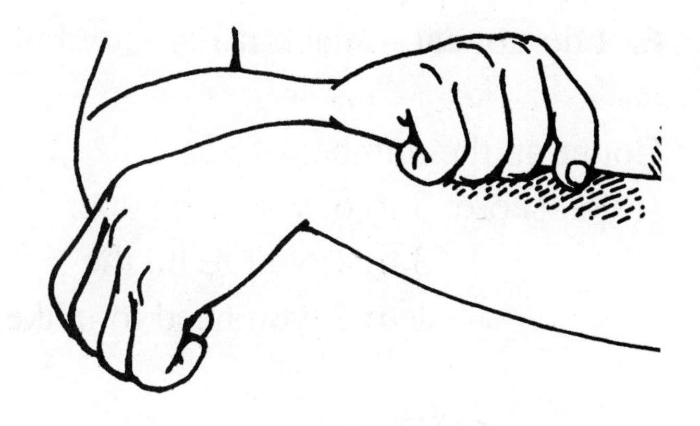

Name ______________________________ Date ______________

SNACKING TEST

Take this test about snacking. Check *Yes* or *No* next to each question.

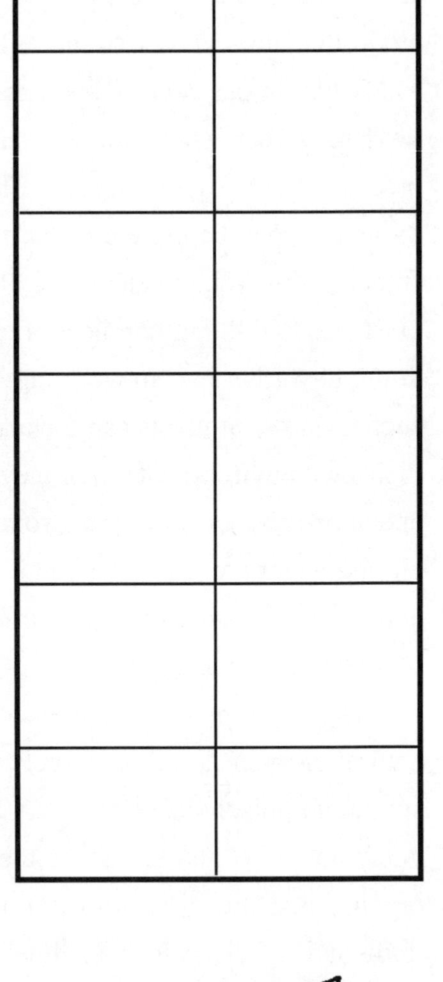

YES	NO

1. I do not snack if it is close to mealtime. This can ruin my appetite.

2. I brush my teeth or rinse my mouth after snacking. This helps prevent cavities.

3. I do not snack on sticky, sweet food. These can stick to my teeth and cause cavities.

4. I snack on healthful foods such as raw vegetables, fresh fruit, nuts, cheese, or even pizza.

5. I do not snack on junk foods such as soda pop, potato chips, pretzels, cookies, and candy.

6. I do not eat a snack if I am not hungry.

Count up the number of yes's.
If you score: 5 to 6, you're a super snacker!
3 to 4, you're on the right track.
1 to 2, you need to make a few changes.

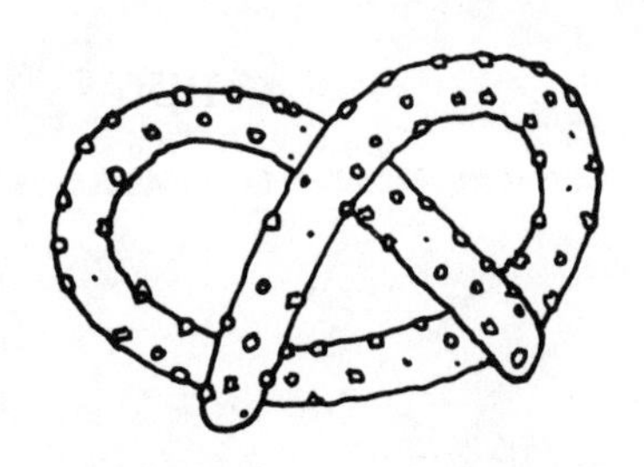

Name ____________________ Date ____________

Comparing Food Groups

You are waiting for the lunch bell to ring. Why are you so hungry? Well, you had a busy day. You walked to school, hurrying so that you wouldn't be late. Those exercises in gym class were fun, but tiring. You also worked hard in your other classes today. You used up a lot of energy. Now, there's only one way to get all the energy back. You must eat food. No wonder you are looking forward to lunch.

Look at the food pyramid. It shows the five basic food groups. Foods in the same food group provide about the same nutrients for your body. Did you know that nutrients are substances in food that you need so that you can grow and have energy?

Your diet is the combination of foods you eat each day. A *balanced diet* gives you the right amount of food each day from the five basic food groups. After you have made your food choices from the five basic food groups, you might make additional choices. You might decide to add items that include fats and oils from the top of the pyramid. Foods that have fats and oils include cakes, cookies, and french fries. Eating too much of them can cause you to gain extra weight and may be harmful to your teeth.

Scientists and doctors continue to do research about food and share their findings. More than ever before, people are concerned about what they eat. They are finding out that a balanced diet can be an important part of good health. In the past, nutrition experts put foods in four food groups. Compare the new food pyramid above with the old food groups next to it.

Go on to the next page.

Name ________________________________ Date ________________

COMPARING FOOD GROUPS, P. 2

Answer these questions.

1. How are the two ways to group foods the same?

____

____

____

2. How are they different?

____

____

____

____

3. Why do you think the number of groups has changed? ______________

____

____

4. Why is the food pyramid more helpful than the old food groups when you're planning what to eat? ______________________

____

____

5. Use the food pyramid to plan a healthful after-school snack. You may combine items from two or more food groups to make your snack. For example, you could eat yogurt with fruit. Write down your idea for a healthful snack.

____

____

____

Name ______________________________ Date ______________

WHAT NUTRIENTS DOES YOUR BODY NEED?

Food contains **nutrients** that your body needs. There are many different kinds of nutrients. Nutrients that give you energy are called carbohydrates. Nutrients that help to build your muscles are called proteins. Nutrients that protect your body organs are called fats. Fats are also a good source of energy. Minerals are nutrients that help make up some of your body parts. One mineral is calcium. Calcium is needed to build bones. Another mineral is iron. Iron is needed to make blood. Vitamins are used in different ways. For instance, vitamin A helps your eyesight. Vitamin C helps fight germs in the body. The chart below lists some nutrients. The chart also gives food sources for these nutrients.

NUTRIENT	FOOD SOURCE
Carbohydrate	Potato, sugar, bread, cereals, pasta
Protein	Fish, chicken, eggs, beans, peas
Fats	Oil, lard, margarine, butter, nuts
Vitamin A	Carrots, apricots
Vitamin C	Tomatoes, oranges, strawberries
Calcium	Milk, cheese, yogurt
Iron	Spinach, liver

Go on to the next page.

Name ______________________________ Date ______________

WHAT NUTRIENTS DOES YOUR BODY NEED?, P. 2

Complete this nutrient puzzle.

ACROSS

2. These foods give you energy.
3. This vegetable is a good source of vitamin C.
6. These nutrients help your body fight diseases.
7. This food is a good source of protein.
8. This vegetable is a good source of iron.

DOWN

1. This nutrient helps build strong muscles.
4. This food is a good source of calcium.
5. This food is in the fats group.
7. Margarine is a good source of this nutrient.

Name ______________________ Date ____________

IT'S IN THERE

Look at the foods on the grocery shelves below. They all contain nutrients. Write the name of each food under the right nutrient group.

CARBOHYDRATES **FATS** **VITAMIN C**

PROTEIN **VITAMIN A** **CALCIUM**

Name ______________________________ Date ______________

FAT TEST

Do you know what foods contain fats? Fats are found in a lot of different foods. Sometimes you can't see fats. For example, there is fat in milk, meat, and eggs.

You can do this test for fats:

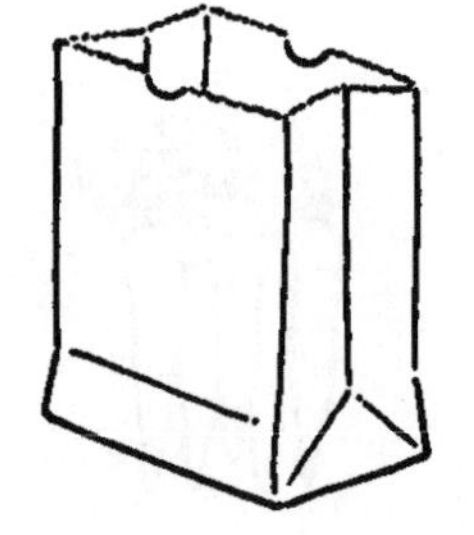

1. Get a brown paper bag, a pair of scissors, and some different foods to test.
2. Cut the paper bag into 3" squares. Write the name of each food you wish to test on a different square.
3. Rub a piece of food on a square until it leaves a wet spot. If the food is liquid, put a drop of it on the square.

4. Set aside the squares to dry.
5. When the squares are dry, hold them up to the light. If there is a greasy spot, the food contains fat. Fill in the chart below.

FOOD	FAT	NO FAT
Milk		

Name ______________________________ Date ______________

PREVENTING CAVITIES

Your permanent teeth began to appear when you were about 6 years old. You will have most of them by the time you are 12 or 13. You want your teeth to be healthy and strong. You don't want to have cavities.

Cavities are caused by bacteria that are always in your mouth. These bacteria feed on the food that is stuck in your teeth. They form an acid that eats holes in your teeth. You can help prevent cavities.

Here are some rules for teeth care.

1. Brush your teeth every morning and evening.
2. Use a toothbrush that has soft bristles and a flat top.
3. Visit the dentist twice a year.
4. Do not eat a lot of sweets (this includes chewing gum and soda).
5. Eat fruits and dairy products every day.
6. After eating, brush your teeth or at least rinse your mouth.

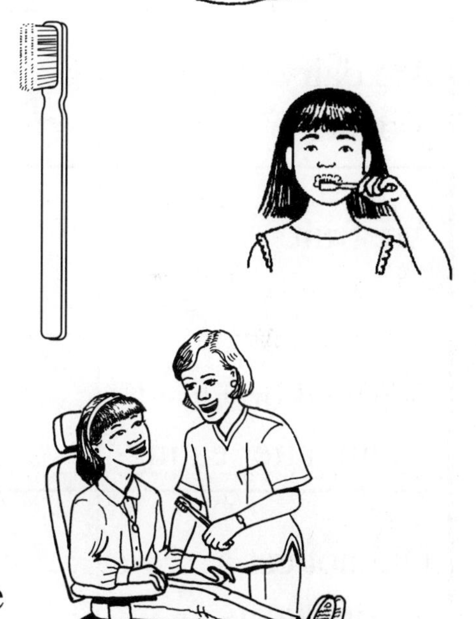

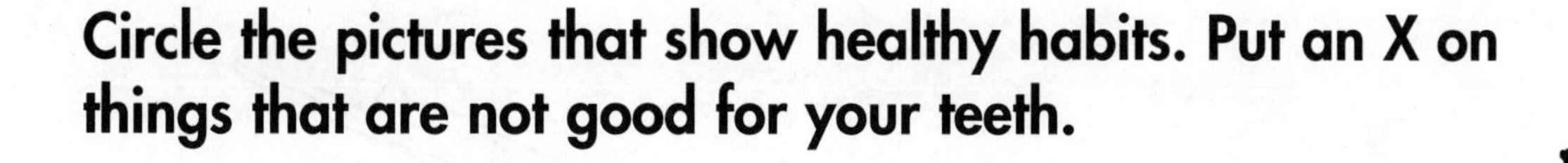

Circle the pictures that show healthy habits. Put an X on things that are not good for your teeth.

Go on to the next page.

Name ______________________ Date ______________

PREVENTING CAVITIES, P. 2

Do you follow all these habits? Keep this chart for one week. At the end of each day, check off the habits you followed that day.

	Mon.	Tues.	Wed.	Thurs.	Fri.	Sat.	Sun.
Brushed my teeth in the morning							
Brushed my teeth in the evening							
Ate dairy products							
Ate fruit							
Brushed or rinsed mouth after eating							
Did not eat a lot of sweets							

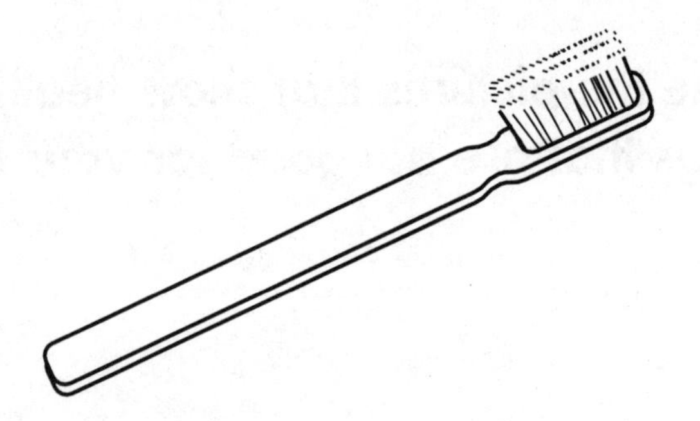

Name ________________________________ Date ________________

HYGIENE HABITS

Bacteria can cause diseases. It is important to follow good hygiene habits to prevent their spread. Do you follow all these habits? Check *Yes* or *No*.

	YES	NO
1. I carry tissues and cover my nose and mouth when I sneeze.		
2. I wash my hands after using the bathroom.		
3. I do not drink or swim in polluted water.		
4. I wash my hands before handling food.		
5. I do not drink out of glasses or use utensils that others have used unless they are washed first.		

Draw an X over the people who are not following good habits.

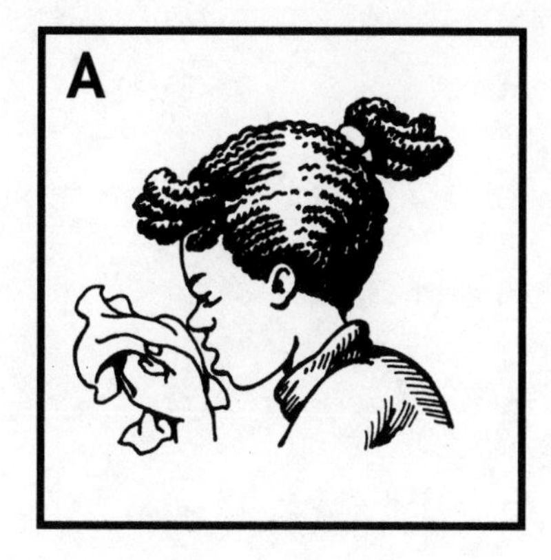

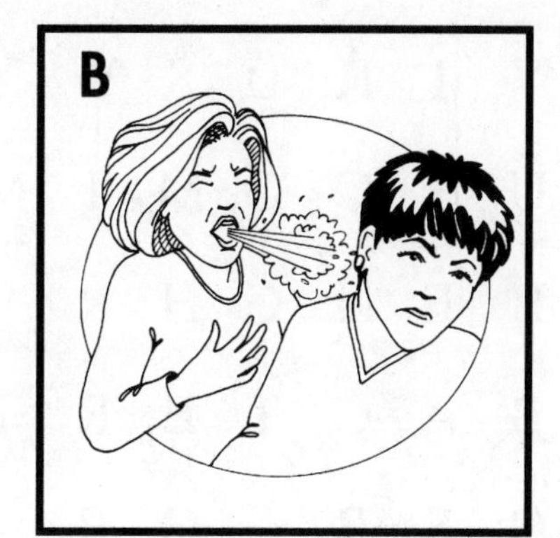

Name ______________________________ Date ______________

Germs That Cause Diseases

Diseases are spread from one person to another by germs that we cannot see. When germs enter your body, they multiply and cause disease.

What germs can make you ill? Bacteria are germs that are on everything you touch and in the water you drink. Many diseases, such as tuberculosis and blood poisoning, are caused by germs. Another kind of germ that can make you ill is called a virus. A virus is the smallest known disease-causing germ. Some viruses cause colds. A fungus is another germ that can make you ill. Fungi are usually found in damp places. Athlete's foot and ringworm are caused by fungi.

You can help protect yourself and others from diseases by getting vaccinations and by following good health practices.

Locate the following words in the Word Search below. These words may be down, across, or on the diagonal. Then, on another sheet of paper, use each word in a sentence.

disease	**germ**	**microbe**	**fungus**
bacteria	**virus**	**vaccine**	**health**

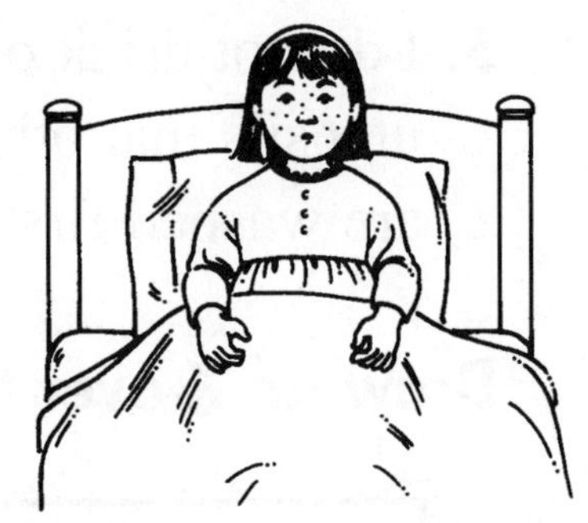

F N V C D E K F D L E K J W J

V U A Z D Y J V I R U S W X A

P O C F U N G U S C O M I M L

M I C R O B E B E N G H R V U

S T I R G I H B A C T E R I A

A Y N M A B A O S P G Q R T S

X S E T L P Q H E A L T H M N

Name ______________________________ Date ______________

Making Yogurt

Not all bacteria are harmful. For example, many different dairy products are made by adding bacteria to milk. Bacteria grow in warm milk and change it. Cheese, yogurt, and buttermilk are made in this way.

You can make your own yogurt at home.

A. Gather the following:
1 container of plain yogurt, 1 quart of milk, a measuring cup, a candy thermometer, a saucepan, a clean glass container that will hold 1 quart of liquid, and a towel.

B. Warm the milk in a pan to 71°C (160°F). This will kill any bacteria that will spoil the milk.

C. Cool the milk to 43°C (110°F).

D. Mix 1/2 cup of yogurt with 1/2 cup of warm milk. Mix thoroughly.

E. Add the remainder of the plain yogurt and stir.

F. Add this to the rest of the milk in the pan. Stir.

G. When the two are thoroughly mixed, place in the glass container. To keep in the heat, wrap the container in a towel. Put the container in an oven (not turned on). Let stand for ten hours. To stop the bacterial growth, refrigerate the yogurt.

H. There are many ways to enjoy yogurt. Try mixing some with diced fresh fruit, honey, and walnuts.

Name ______________________ Date ____________

USING BACTERIA TO STAY HEALTHY

Have you ever had a vaccination? Did you ever drink polio vaccine? Doctors put bacteria into your body to keep you healthy. They use a weakened form of bacteria. The bacteria are too weak to grow in your body and make you sick. But the weakened bacteria make your body change. They cause your body to make *antibodies.* If the same kind of bacteria enter your body again, these antibodies protect you.

Look at the chart. It tells what vaccinations many doctors suggest. Use the chart to answer the questions.

AGE	TYPE OF VACCINATION
2 months	diphtheria, whooping cough (pertussis), tetanus, [d.p.t.]; polio
4 months	d.p.t.; polio
6 months	d.p.t.
15 months	German measles, measles, mumps
18 months	d.p.t. booster; polio booster
4-6 years	d.p.t. booster; polio booster
14-16 years	d.p.t. booster

1. At what age are vaccinations first given? ____________

2. What vaccinations are given at 6 months? ____________

____

3. What vaccinations are given at 15 months? ____________

____

Ask your parents what vaccinations you have had. Make a list of them. At what age did you receive them?

Name ______________________ Date ______________

Measuring Height

Look around at your classmates. Notice that they are not all the same height. There are several reasons for this. Everyone is born with a set of traits he or she received from his or her parents. If both your parents are tall, you will probably also be tall. If both are short, you will probably be short. If one of your parents is tall and one short, you may be medium height. Another reason for the difference in your classmates' heights is that people have different growth patterns. One person may grow faster in the beginning, yet end up shorter than someone else who started later. It may take you longer to reach the same height as one of your classmates.

This bar graph shows the growth of one student from age 1 to 10.

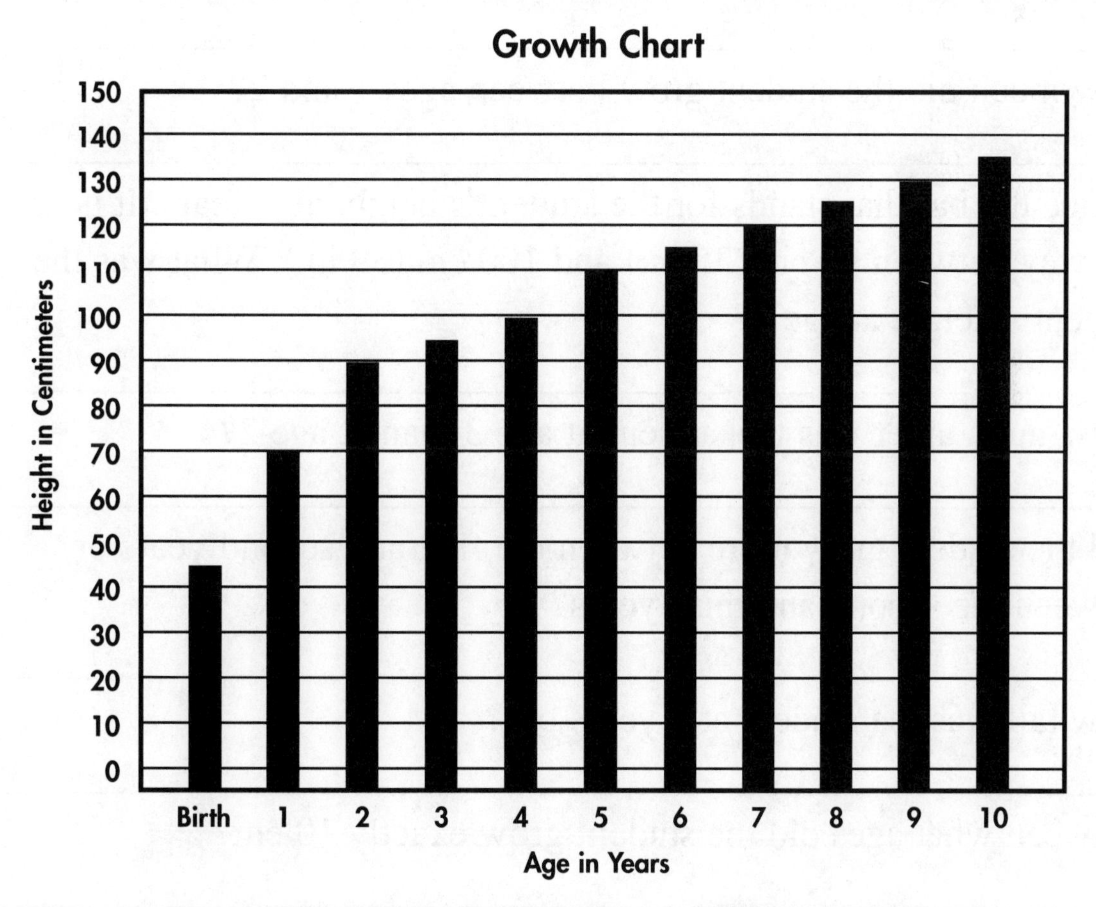

A bar graph shows information about something and helps you compare numbers or amounts. Each bar in this graph represents the number of centimeters tall that the student was at a certain age. As you can see, the student grew taller each year.

Go on to the next page.

Name ______________________________ Date ______________

MEASURING HEIGHT, P. 2

Suppose you want to know how tall the student was at the age of one. Find "Age in Years" at the bottom of the graph. The numbers at the bottom show age from 1 to 10 years. The numbers along the left side stand for different heights from 0 to 150 centimeters (0-59 inches). To find the student's height at one year, find the bar marked "1 year." The top of the bar is the same height as 70 cm (27 in.). So the student was 70 cm (27 in.) tall at one year of age.

1. How tall was the student at 2 years of age?

____

2. How much did the student grow between ages 1 and 2?

____

3. Notice the bar that stands for the student's height at 3 years. It is about halfway between 90 cm (35 in.) and 100 cm (39 in.). What was the student's height at age 3?

____

4. How much taller was the student at age 3 than at age 2?

____

5. Did the student grow more between the first and second years or between the second and third years?

____

6. How tall was the student at 7 years old?

____

7. Between what ages did the student grow exactly 10 cm?

____

8. During which two years did the student grow the most?

____

9. How many centimeters did the student grow in 10 years?

____

Name ______________________ Date ______________

EXERCISE

Exercise is important for everyone. If you do not use your muscles, you will become weak. Even astronauts must exercise. Since their bodies are not working against the full force of gravity, they hardly use their muscles in space. Bones and muscles get weak when there is not much gravity.

In 1970, three Soviet cosmonauts, who had been in space for 18 days, had to be carried from their spacecraft because their muscles were too weak to move their bones. Now astronauts exercise in space. They use a treadmill on which they run in place against a moving track. Using the treadmill makes their arm and leg muscles work harder. By walking the treadmill for 15 to 30 minutes each day, they keep their bones and muscles healthy.

Working out on the treadmill is a form of aerobic exercise. Aerobic exercise forces the body to use a large amount of oxygen over a long period of time. If you are in good health, you should do some form of aerobic exercise at least three times a week for at least 15 minutes without stopping.

Answer these questions.

1. Why do astronauts need to exercise in space?

____

____

____

Go on to the next page.

Name ______________________ Date ____________

EXERCISE, P. 2

2. How does the treadmill help astronauts?

____

____

Here's an exercise for you to try:
Place your hands against the wall at chest level. Try running in place while pushing for 5 minutes.

3. How is this exercise like running on a treadmill?

____

____

____

Here is an exercise chart for some students. The chart shows how the students' heartbeats changed after exercising for 5 minutes. Use the chart to answer the questions.

Exercise Chart		
Student	**Heartbeats Before**	**Heartbeats After**
Tom	80	134
Lynn	75	130
Carlos	68	124

4. Who had the lowest number of heartbeats before exercising?

____

5. Who had the lowest number of heartbeats after exercising?

____

6. Whose heartbeats changed the most because of exercise?

____

Name ______________________________ Date ______________

MUSCLE MAGIC

Sometimes your muscle cells keep working even after you tell them to stop.

Try this trick.

A. Stand in a doorway. Press your arms upward against the door frame as hard as you can. Do this for 30 seconds. You could also have a partner hold your arms down while you press them up.

B. Relax your arms. What happens? Your arms will rise even after you tell them to stop!

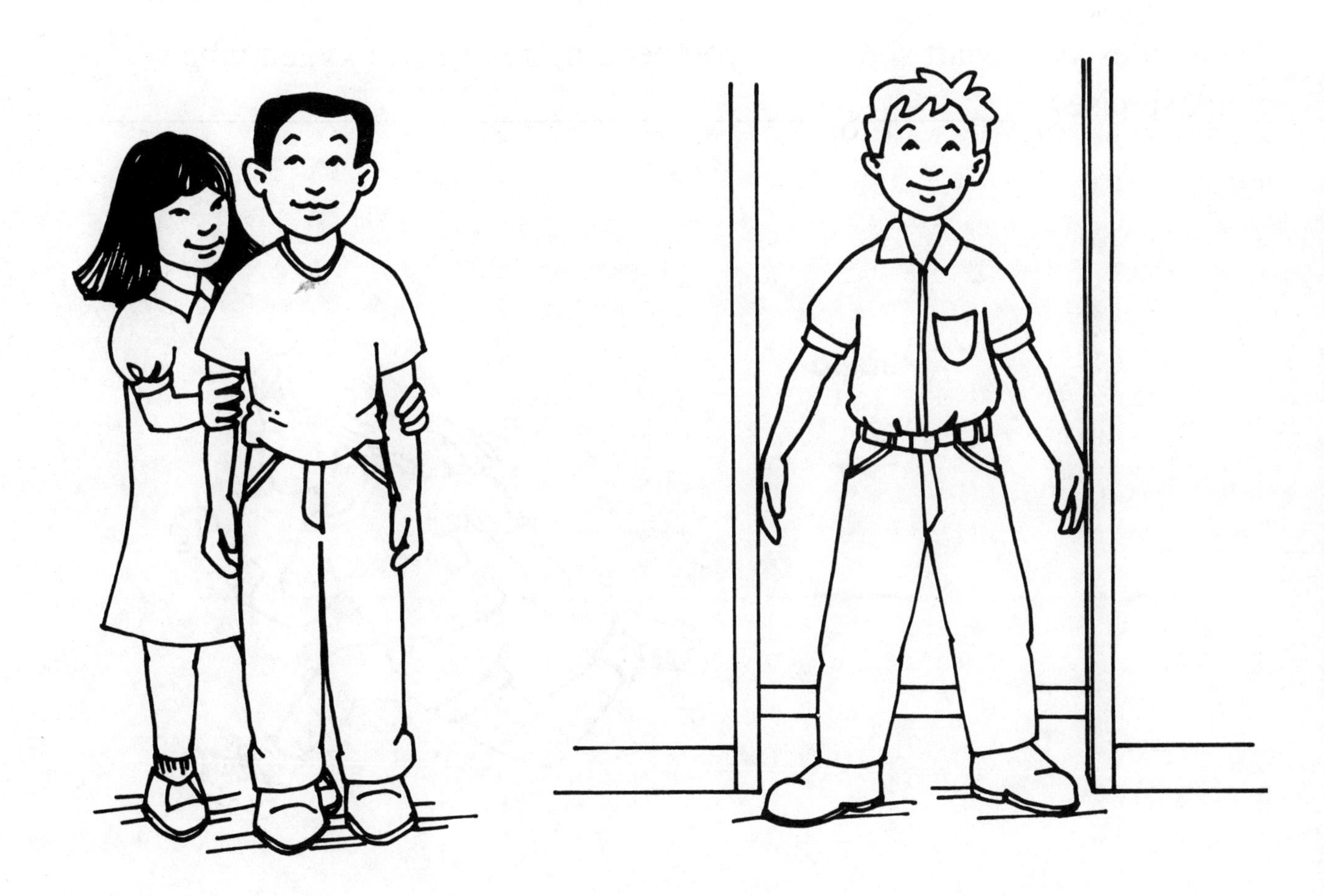

Name ________________________________ Date ________________

How Much Air Do You Need?

Sit in a chair and relax. Using a watch with a second hand, count the number of times you breathe in during one minute. Write the number in the chart below.

Now run in place for one minute. Then sit down. Count the number of times you breathe in for one minute. Fill in the chart below.

ACTIVITY	NUMBER OF IN-BREATHS/MINUTE
Sitting	
Running	

What does your chart show? Do you need more or less oxygen when you exercise? ________________________________

Name ______________________ Date ______________

POLLEN IN THE AIR

Do you start sneezing a lot in spring and summer? Do your eyes get red, and does your nose run? Many people who suffer this way are allergic to pollen. They have hay fever.

Plants use pollen to reproduce. Pollen is made in flowers. In the spring, many trees have flowers that make lots of pollen. All summer, grasses in lawns and fields produce pollen. Many people are allergic to the pollen produced by ragweed, which flowers in the early fall. Pollen grains are very light and are carried long distances by the wind.

When there is a lot of pollen in the air, people who are allergic to pollen become very uncomfortable. People who watch air quality count the number of pollen grains in the air. They do this every day. A sample of air is taken, and the number of pollen grains in the sample is counted. Then the results are given to newspapers and television stations, which report the number as the pollen index for the day. The actual amount of pollen in the air can vary quite a bit throughout one day, as winds shift and rain falls.

The graph below shows the pollen counts for one day each month over one year in Santa Barbara, California. Use the graph to answer the questions on the next page.

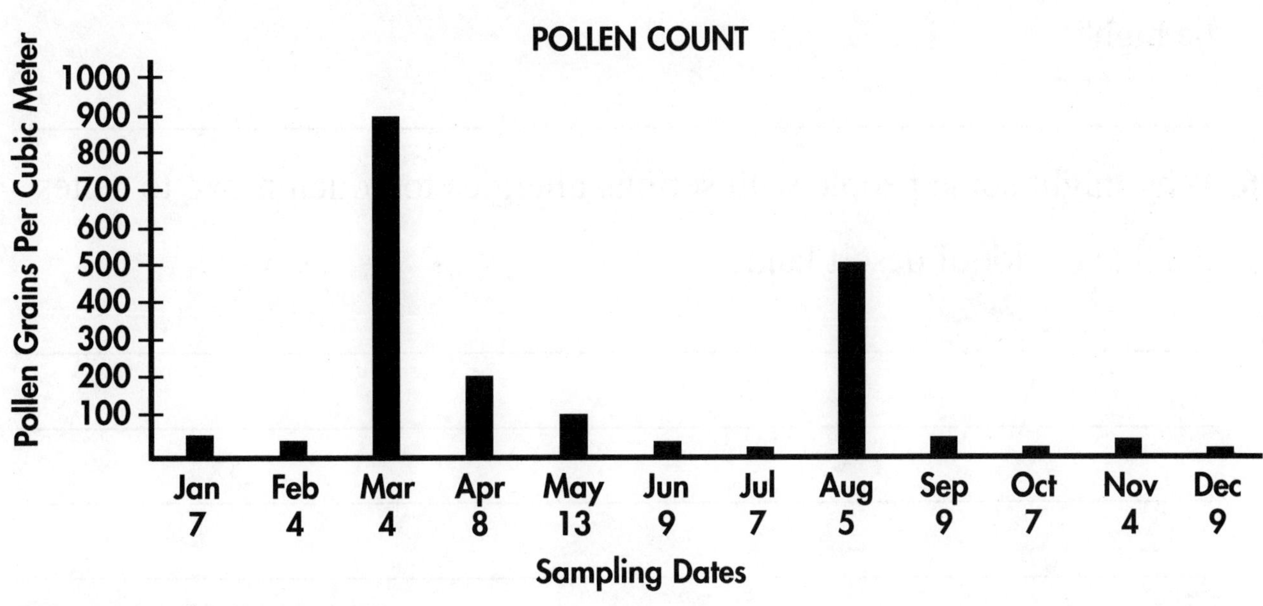

Go on to the next page.

Name ______________________ Date ______________

Pollen in the Air, p. 2

Answer these questions.

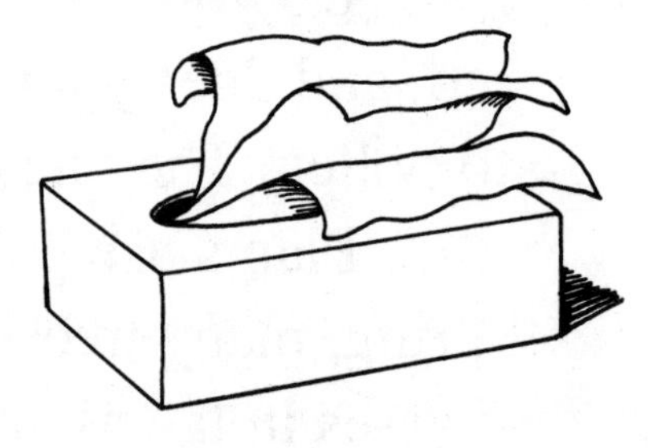

1. How many pollen grains were present in a cubic meter of air on April 8?

____

2. How many pollen grains were there in a cubic meter on August 5?

____

3. If you had hay fever, on which day shown on the graph would you have been most uncomfortable?

____

4. What do you think happens to the number of pollen grains in the air when it is raining?

____

5. How could the pollen count in a city with few trees and weeds still be high?

____

6. Why might some people with serious allergies to pollen move to states that have a lot of desert land?

____

____

____

____

Name ______________________________ Date ______________

SMOG DETECTOR

One form of air pollution is called smog. It is made up of smoke that is trapped near the Earth. Smog can irritate the eyes, nose, throat, and lungs. It can also make it hard for some people to breathe.

In the picture below, put an X over the area that you think might cause smog. Then draw three things that cause air pollution in your community.

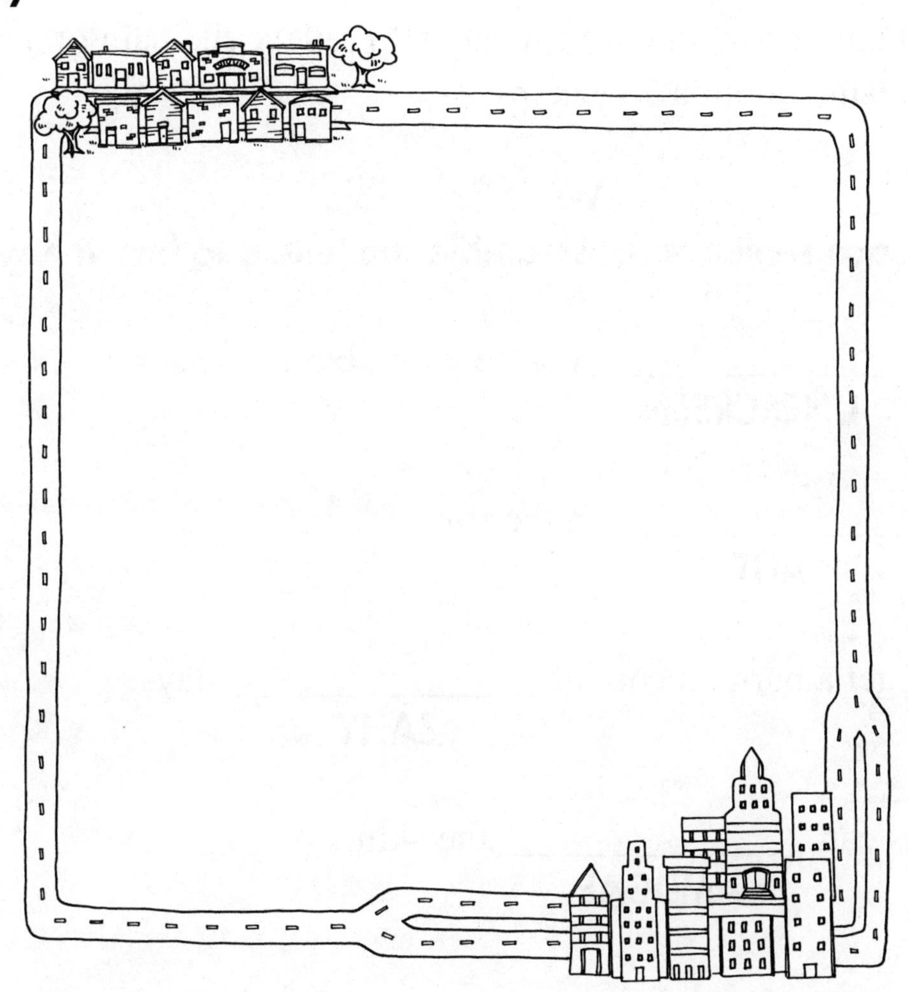

What can you do to help prevent air pollution in your community?

____

____

Name ______________________ Date ______________

SUNBURNS

Have you ever had a sunburn? Some people burn more easily than others, but no one is immune. A sunburn can be very painful. It is wise to avoid sunburns. They can damage the skin.

If you know you will be exposed to the sun for many hours, be prepared! Remember that you can get a burn even on hazy or cool days. Keep yourself covered. Wear a hat. Apply a sunscreen before going out. Reapply every two hours or after swimming.

Cold compresses and aspirin can help relieve the pain of a severe sunburn. Calamine lotion also helps.

Word Scramble

Complete each sentence. Unscramble the letters to find the words.

1. Apply a ______________ before going out in the sun.
UNESCRESN

2. Wear a ______________ in the sun.
AHT

3. You can get a burn on cool or ______________ days.
ZAHY

4. Sunburns can ______________ the skin.
AMDGAE

5. Cold compresses and ______________ can help relieve the pain of sunburn.
PSIAINR

Name ______________________________ Date ______________

SMOKING

Is smoking glamorous? Does smoking make you look grown-up? Do good athletes smoke? The answer to all these questions is no.

Smoking damages the body in many ways. People who smoke are sick more often than people who don't. They have more cases of cancer and heart disease. They even have more colds. Smoking also hurts other people who do not smoke.

Cut out some of the cigarette ads from magazines or newspapers. Or you might want to draw your own ads. Paste them in the space below and write the message that you think the ad is trying to tell you. Then write an ad or make a poster that tells the truth about smoking.

Name ______________________________ Date ______________

INSECT PESTS

Do you know these insects? They are common pests. There are things you can do to outwit them when you are outdoors.

Mosquitoes: Keep cool and bathe often. Wear light colors. Wear long sleeves and pants in the evening and early morning. An ice cube or calamine lotion can relieve itching.

Horseflies: Watch out on warm, sunny days! Wear light colors. Wash fly bites well.

Black flies: Black flies like wooded areas near water. They are active during the day. Wear light colors.

Honeybees and Wasps: Wear light colors. Avoid flowery prints. Do not wear hair spray or perfume. Do not walk barefoot. Do not wave arms at bees and wasps. If stung, apply ice cube and baking soda paste. If you have severe local swelling or any other allergic reaction, get to the doctor at once.

Get a book on insects. Find the names of these insects. Label them.

A

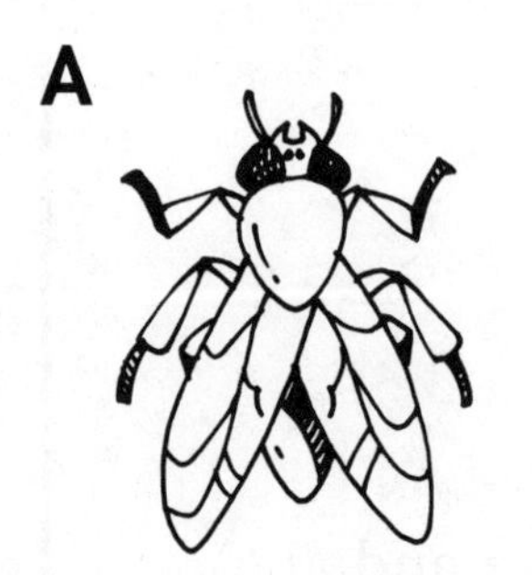

____

B

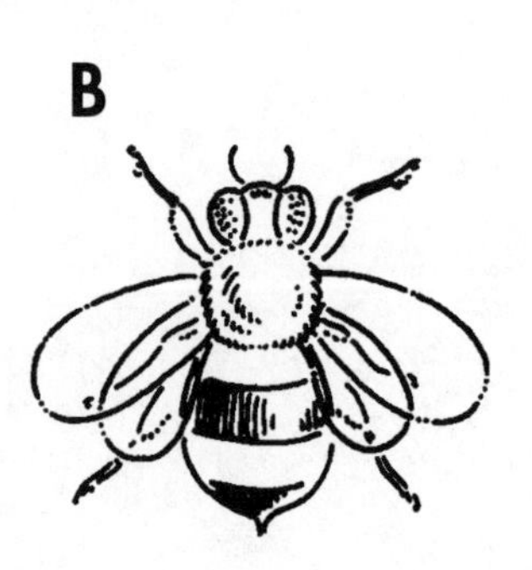

____

C

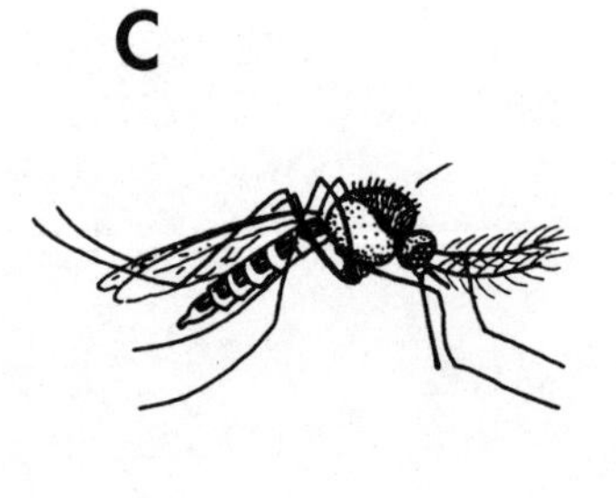

____

D

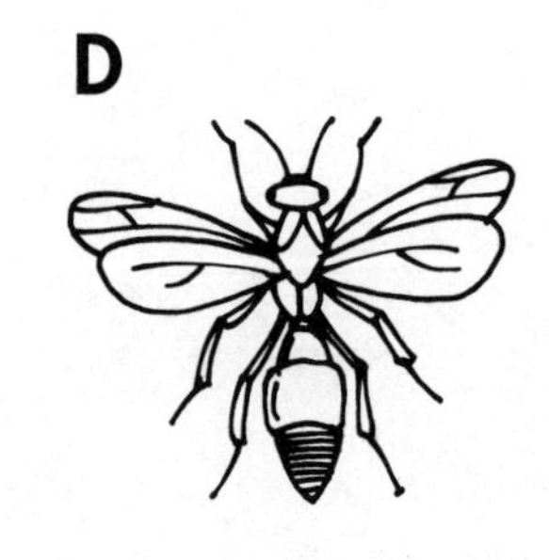

____

Name ______________________________ Date ______________

Lakes, Ponds, Mosquitoes, and Disease

You might think that mosquitoes read travel advertisements. They seem to appear at all the lakes and ponds as soon as the weather is warm. Mosquitoes do not visit these places for pleasure. They are born there.

Mosquitoes lay their eggs in fresh water. The water can be in a quiet corner of a pond, in a ditch, or in an old tire. A puddle will do, too. When the eggs hatch, the young don't look like mosquitoes at all. In this stage of their life cycle, the young are called *wrigglers*, and they look like tiny caterpillars. Wrigglers can't fly like adult mosquitoes. They breathe through a tiny tube like a snorkel that sticks out of the water. Wrigglers find many tiny organisms to feed on in the water.

Answer these questions.

1. Why are so many mosquitoes found near ponds and lakes?

____

2. What does a mosquito look like when it first comes out of the egg?

____

3. What do young mosquitoes eat?

____

After a time. wrigglers go into a resting stage, and then they become adults. All this happens in a few days. As adults, mosquitoes can fly. The females get their nourishment by sipping the blood of people and animals. Have you ever been bitten by a mosquito?

When mosquitoes bite you, you may get sick. Mosquitoes do not cause disease. They just carry—from one person to another—the organisms that cause the disease. The drawing on page 136 shows the sharp mouth parts that allow the mosquito to cut into the skin. Look at the chart on page 136, which shows some diseases that mosquitoes carry from one person to another.

Go on to the next page.

Name ______________________ Date ____________

LAKES, PONDS, MOSQUITOES, AND DISEASE, P. 2

DISEASES SPREAD BY MOSQUITOES
Yellow Fever Malaria Encephalitis

4. How are diseases carried from person to person by mosquitoes?

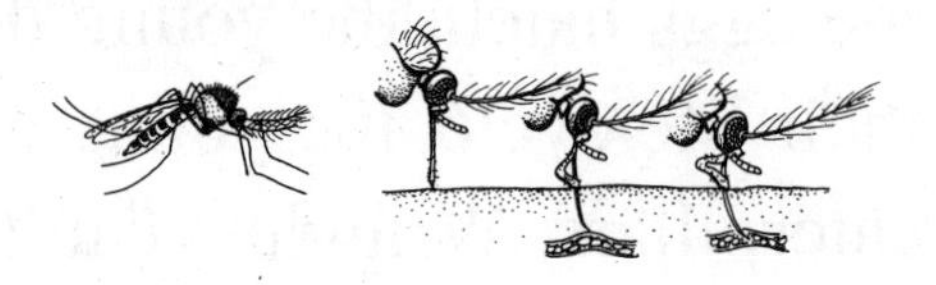

____

____

____

5. What feature does a mosquito have that allows it to pierce the skin?

____

____

People try to control mosquitoes by spraying chemicals to kill them. These chemicals may harm birds and other animals. A safer way to control mosquitoes is to drain ponds and puddles where they lay their eggs. However, this may be harmful to other wildlife that lives in ponds. Other ways to control the pests include encouraging natural enemies, like certain flies, bats, or fish, or spreading diseases that kill mosquitoes.

6. Why is the practice of draining ponds to kill mosquitoes not always a good idea?

____

7. What methods for controlling mosquitoes might be better for the environment?

____

Name ______________________________ Date ______________

HIKERS, BEWARE!

Many people get a rash from plants found outdoors. The most common plant to cause a rash is poison ivy. The sap in poison ivy causes a rash hours or days after touching it. You can also get a rash from clothes or pets that have touched the plant. Even smoke from burning poison ivy can cause a rash.

There are some things you can do if you touch poison ivy. Wash your skin and clothes with soap and water at once.

If you get poison ivy, it will last one or two weeks. It goes away even if not treated. Scratching the rash does not spread it as many people think. You can buy calamine lotion at the drugstore to help stop the itching.

What does poison ivy look like? It is a woody vine or small bush. The leaves are bright green and grow three to a stem. In fall, it turns orange or red. The plant has small, greenish-white flowers. Later, it gets small, white berries.

There are two other plants that cause a rash. They are poison oak

A ______________________

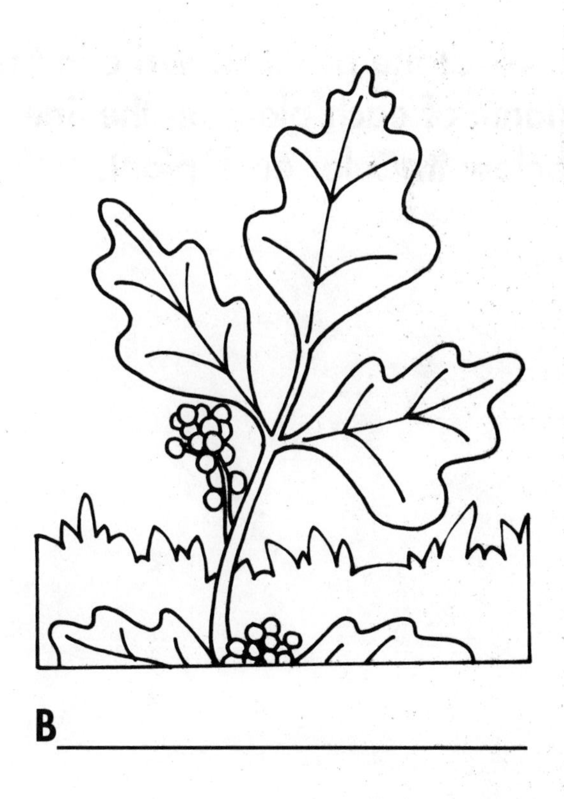

B ______________________

Go on to the next page.

Name ______________________________ Date ______________

Hikers, Beware!, p. 2

and poison sumac. Poison oak also has three leaves on a stem, but its leaves are rounded. The leaves have short, soft hairs on the bottom. Poison oak has small, yellow flowers and small, white berries. It grows as a small shrub about three-feet tall.

Poison sumac has shiny, green leaves that grow 7 to 13 on a stem. It can be a small tree or a large bush. The stems are often reddish. Poison sumac has small, greenish flowers with gray berries.

C______________________

Look at the pictures. Write in the name of each plant on the line below it. Color each plant.

Name ________________________________ Date ________________

Poison Plants

Fill in the blanks in the following sentences about poison plants. Find each word in the Word Search. Words can be up, down, across, or on the diagonal.

1. You need to watch for poison plants when you are ______________.
2. ______________ ______________ is the best-known poison plant and grows as a woody vine or a small bush.
3. ______________ lotion can relieve the itch of poison ivy.
4. A small tree with 7 to 13 leaves on a stem is called ______________ ______________.
5. ______________ ______________ has three leaves on a stem and the leaves are rounded.
6. There are three types of poison ______________.
7. The ______________ from a poison plant causes a rash.
8. A ______________ from poison ivy is not spread by scratching.
9. All three poison plants have ______________.

L	P	O	S	G	A	M	R	E	L	D
P	O	I	S	O	N	S	U	M	A	C
L	I	U	B	G	F	A	D	P	L	A
A	S	C	T	I	O	P	E	A	L	L
N	O	H	I	D	M	U	R	O	E	A
T	N	A	L	Q	O	A	E	M	R	M
S	O	T	G	P	S	O	E	A	G	I
V	A	J	O	H	K	Z	R	O	I	N
R	K	A	L	S	J	N	K	S	C	E
B	E	R	R	I	E	S	D	F	C	M
O	S	Y	V	I	N	O	S	I	O	P

Name ______________________________ Date ______________

Poisons in Your Home

Did you know that there might be poisons in your home? Many of the things adults use to clean and care for their homes are poisonous. Children should never use these things unless they are with an adult. People should always wash their hands carefully after using poisons. They should be kept in a safe place where children cannot get near them.

Talk to an adult in your home about the poisons there. Discuss which things are poisonous. Talk about where these things are kept and whether these places are safe.

1. What poisons do you have in your home? ______________________________

____

____

2. What are the poisons used for? ______________________________

____

____

3. Where are these poisons kept? ______________________________

____

____

4. What has been done to be sure children do not use the poisons in your home? ______________________________

____

____

Name ______________________ Date ______________

HEALTHY HABITS

Keep a record for one week of all your activities. Fill in the chart at the end of each day. What foods did you eat? Did you exercise? What did you do to keep your body and your teeth clean? Think about anything you may have done to keep yourself safe from hazards in your environment, or ways that you helped your environment. Did you make good choices? Do you have healthy habits?

Monday	Tuesday	Wednesday	Thursday	Friday

Name ______________________________ Date ________________

UNIT 4 SCIENCE FAIR IDEAS

A science fair project can help you to understand the world around you better. Choose a topic that interests you. Then use the scientific method to develop your project. Here is an example:

1. **PROBLEM:** How does activity affect a person's pulse rate?
2. **HYPOTHESIS:** When a person is active for an extended period of time, his or her pulse rate will become faster. When a person is inactive, his or her pulse rate stays slow.
3. **EXPERIMENTATION:** Do several experiments with several different people of the same age group. Take each person's pulse before and after each activity. Experiment with different lengths of time. For example, for a running in place activity, you might see how one minute of running affects the pulse rate. After letting the person rest, see if two minutes affects them differently. Keep careful records of all your findings.
4. **OBSERVATION:** When people are more active, they breathe faster, they have more color in their faces, and their pulse rates go higher. When they are inactive, they are paler, and their breathing is more regular.
5. **CONCLUSION:** Exercise raises the pulse rate. The more a person exercises, the higher the pulse rate goes. If a person is inactive, the pulse rate is slow.
6. **COMPARISON:** Conclusion agrees with hypothesis.
7. **PRESENTATION:** Display your records and any pictures or drawings you did with your observations. You may want to show your findings on a chart or graph.
8. **RESOURCES:** Tell of any reading you did to help you with your experiment. Tell who helped you to get materials or set up your experiment.

Other Project Ideas

1. Why do human babies stay with their parents longer than animal babies do?
2. How do the foods people eat affect the way they look and feel?
3. What happens if people do not get enough sleep?
4. How does exercise make our bodies stronger?

Life Science
Grade Three
Answer Key

P. 9 Unit 1 Assessment: 1. non-living, 2. food, 3. move, 4. leaves, 5. reproduce, 6. cells, 7. wall, 8. membrane, 9. nucleus, 10. reproduction, 11. cytoplasm

P. 10 Unit 2 Assessment: ACROSS: 3. chlorophyll, 6. pollen, 7. fungi, 8. stamen, 9. cones, DOWN: 1. pistil, 2. ovule, 4. pollinate, 5. spores

P. 11 Unit 3 Assessment: 1. traits, 2. backbones, 3. Amphibians, 4. mammals, 5. teeth, 6. beaks, 7. adapt, 8. instincts, 9. camouflage, 10. migration

P. 12 Unit 4 Assessment: 1. grains, 2. germs, 3. dentist, 4. pollen, 5. ivy, 6. sunburn, 7. mineral, 8. smog, 9. sneeze, 10. teeth
CHALLENGE: Good Hygiene

P. 16 1. No, the tree is alive but the log is not because the log can not grow, does not need oxygen or food, and can not reproduce. The tree can do all of these., 2. The tree needs food, water, and air., 3. The bird needs food, water, and air., 4. No. The rock is not alive., 5. The rock is like the log because the log is now dead. It is no longer growing or alive, so it does not need the things that a living thing needs, and it cannot reproduce., 6. Yes, the bird is alive and living things reproduce.

P. 17 Circle the dog, the bear, the grass, the snake, and the girl. Put a square around *moves by itself, reproduces, needs food,* and *grows.*

P. 18 a. cell wall, b. membrane, c. cytoplasm, d. nucleus

P. 19 1. cell wall, 2. cell membrane, 3. cytoplasm, 4. nucleus; Students draw cell and label parts correctly. (Use cell on p. 18 as a guide.)

P. 20 a. nucleus, b. cytoplasm, c. cell membrane, D. 1. skinlike covering that surrounds the cell, 2. living liquid inside a cell, 3. control center of a living cell

P. 21 Students record the results of their experiments.

P. 22 1. the cytoplasm, 2. the membrane, 3. the nucleus, 4. The ball of clay is round and has parts similar to a cell., 5. nucleus, cytoplasm, membrane, cell wall

P. 23 1:00, 2 cells; 2:00, 4 cells; 3:00, 8 cells; 4:00, 16 cells; 5:00, 32 cells; 6:00, 64 cells; 7:00, 128 cells; 8:00, 256 cells; 9:00, 512 cells; 10:00, 1,024 cells; 11:00, 2,048 cells; 12:00, 4,096 cells

P. 24 Drawings should reflect steps 1-4.

P. 25 A. Correct order; 3, 2, 1, 5, 4; The cell takes in extra food. it becomes larger. The nucleus begins to divide. The two parts of the parent cell pull apart and become two new cells., B. Circle the boy.

P. 26 Frog correct order clockwise from top left: 3, 1, 2, 4

P. 27 Under lion: muscle cells, bone cells, skin cells, blood cells; Under cactus: cells with root hairs, cells with chloroplasts, cells with cell walls

P. 28 X's on top left (rod-shaped bacteria), bottom middle (spiral bacteria), and bottom right (round bacteria; Circle around top middle (paramecium), top right (amoeba), and bottom left (euglena)

P. 29 Students record the results of their experiments.

P. 30 Pictures will vary, but should resemble one or more of the protozoans on p. 28

P. 31 The chloroplasts inside the cell trap energy from the sun. Water and carbon dioxide enter the cell through the cell wall. Using energy from the sun, the cell turns the water and the carbon dioxide into food and oxygen.

P. 32 1. 1,000, 2. Days 19, 20, 21, 24, 25, and 26, 3. It stopped growing, 4. a. 1,500, b. 4,000, c. 7,000

P. 33 1. Mold is a simple organism that spoils food., 2. A kind of mold that grows on cloth is mildew., 3. Yeast is a simple organism used to make bread., 4. Mold turns dead wood and leaves into soil.

P. 34 Students record the results of their experiments.

P. 35 1. algae, 2. yeast, 3. mold, 4. mildew

P. 39 Answers may vary. 1. Students should see a small circle of yellow, or less green on the leaf because the chloroplasts in the leaf were not able to get sunlight. 2. The plant will not thrive without oxygen., 3. A plant needs water to live, so it would die., 4. sunlight, oxygen, water

P. 40 Chart: Celery: yes, in tubes, yes; Mushroom: no, spread out in the stem, no

P. 42 1. white birch, 2. redbud and mountain ash, 3. 20 centimeters, 4. crab apple, 5. chokecherry, white birch

P. 43 Observations may vary slightly. Students should observe that the uncovered leaves remain green. They do not soften or dry out. The covered leaves will become pale.

P. 44 1. Possible answers include: It turned pale. it died. Without sunlight it could not make food., 2. No. They remained in sunlight and could make food., 3. The whole plant would probably die.

P. 45 Chart responses will vary. 1. Possible answers include: Both are green. Both have veins and stems., 2. They may differ in color, size, shape, vein patterns, and so on., 3. chlorophyll, 4. Plants need it to make food.

P. 46 2. Answers will vary. One leaf may have more chlorophyll than the other., 3. chlorophyll

P. 47 Students should notice that the leaves of a plant will grow toward the sun. If there has been little sunlight, there may not be a great change.

P. 48 Answers may vary slightly. 1. One of the plants has bubbles coming from it., 2. The one that was in the sun., 3. In the presence of sunlight, the plants carry on photosynthesis, releasing oxygen into the water.

P. 49 Students should label flower correctly; clockwise, from top left: stamen, pistil, petal, ovules, sepal

P. 50 Students should draw 1. an embryo in a seed, 2. a sprout from the seed, 3. a tiny seedling, 4. a growing plant, 5. a flowering plant, 6. a flowering plant with egg cells pointed out

P. 51 If dried beans are used, they should be soaked for an hour for easier opening. Seeds should be opened carefully so that the two halves remain attached. There should be the two fleshy halves of the bean which provide food for the embryo, two first foliage leaves, a small stem-like part that will be the first root, and the place where the halves are attached. Students' drawings should show a bean seed and its embryo.

P. 52 Seed collection charts will vary.

P. 54 1. caves, 2. do not, 3. get food from dead plant and animal material, 4. too wet or too dry, 5. five months, 6. destroys it, 7. can

P. 56 Students' responses will vary.

Pp. 63-64 1. octopus, 2. butterfly, 3. starfish, 4. cricket, 5. crab, 6. caterpillar; students should color the eagle, pigeon, parrot, and bird in nest; students should underline the alligator and lizard; students can X any two of the following: whale, seal, rabbit, squirrel

P. 66 1. Accept reasonable classification selections., 2. Responses will vary. Be sure the reasoning is valid.

Pp. 67-68 Teacher should help students identify each skeleton on p. 67 before they answer the questions on p. 68. (fish, whale, frog, giraffe, bird, gorilla, snake, elephant) 1. Each has a backbone and other bones that support parts of the body., 2. It has no leg bones., 3. It has no fin bones. The whale bones are much bigger., 4. seven, 5. eight, 6. There are no bones in an elephant's trunk., 7. arm bones, 8. The shape of the animal; how large the animal is; how it moves.

P. 69 Line from frog to fly, from lion to zebra, from shark to fish, from squirrel to nuts, from horse to grass

P. 70 M on tiger, dog, snake; P on goat, cow, horse

P. 72 1. chameleon, 2. butterfly, 3. giraffe, 4. starfish, 5. mountain lion, 6. land snail, 7. snake

P. 74 1. Responses will vary., 2. A bird usually eats only one kind of food., 3. hawk, small animals; avocet, insects; thrasher, insects; hummingbird, nectar; pelican, fish

Pp. 75-76 1. an Australian scientist who studied the behavior of geese, 2. in their natural environment and in his home, 3. Instincts are very important in animal behavior., 4. the instinct to follow their mother, 5. Imprinting is the process that occurs when newborn animals attach an instinctive behavior to something they see. Imprinting is an important survival skill., 6. He was the first thing they saw, and he walked slowly and made clucking sounds.

P. 77 1. c, 2. a, 3. b, 4. b

P. 78 Chart: Animal Behaviors points directly to *migration, hibernation,* and *running away.* Migration defends against *cold weather.* Running away defends against *predators.*

P. 80 1. It is very poorly developed and almost helpless. It gets milk and protection from its mother., 2. after five or six months, 3. The young koala eats eucalyptus leaves., 4. A koala is very tiny, has no fur, and is almost helpless. A kitten is much larger, has fur, and is better developed., 5. The kitten drinks milk from its mother off and on. Its teeth begin to appear. It walks around and explores for itself. It starts to eat solid food. The young koala stays attached to a nipple inside its mother's pouch.

P. 81 Students should label cycle: clockwise, from top left, egg, caterpillar, pupa, butterfly

83-84 Frog: lines should be drawn to tongue and color; Fish: lines should be drawn to spots, fins, color, and scales; Heron: lines should be drawn to down feathers, beak, feet, wings, long legs, and contour feathers; Dragonfly: lines should be drawn to wings and body shape; 1. Long legs help the heron wade in the water to look for food., 2. Spots help the trout hide on the pebbly bottom of a stream., 3. The frog's sticky tongue helps it catch insects to eat., 4. The dragonfly uses its wings to fly about looking for food., 5. Fins allow the trout to swim through the water., 6. The heron's long, sharp beak helps it catch fish.

Pp. 85-86 A. Dot-to-dot reveals deer, duck, worm, clam, and fish., B. 1. in water, 2. in the woods, 3. in water, 4. near water, 5. in the ground

P. 87 A. 1. forest, 2. forest, 3. desert, 4. desert, 5. cold north, 6. cold north, 7. farmland and grassland, 8. farmland and grassland, 9. ocean, 10. ocean

P. 88 Answers will vary.

P. 90 1. It blends in with the sand. Its enemies cannot easily see it., 2. It blends in with the grass. Its enemies cannot easily see it., 3. It blends in with the snow. Its enemies cannot easily see it., 5. No. Their colors do not blend in. Their enemies could see them even from a great distance.

P. 91-92 Teacher may explain that the vest or jacket is filled with feathers similar to those of the snowy owl. Point out that the fur of the hat is similar to that of the polar bear. 1. the uncovered one, 2. Answers will vary., 3. The fur keeps the body heat from escaping., 4. The down feathers keep its body heat from escaping., 5. Accept reasonable answers., Chart: Answers will vary.

P. 94 1. T, 2. T, 3. F, 4. T, 5. F, 6. F, 7. T, 8. T

P. 95 Clockwise from top left; 2, 1, 3, 4

P. 97 1. The nest is a safe place for sunfish eggs to hatch., 2. the male, 3. He finds a safe place to build the nest., 4. He fans the water with his fins and tail. A hole forms in the sand., 5. She lays the eggs in the nest. Then she leaves., 6. the male

Pp. 99-100

1. Students may suggest food, water, sunlight, and air., 2. Most students will infer that the caterpillar is in the sac. However, accept all reasonable responses because inferences need not be correct., 3. Most students will realize that the caterpillar metamorphosed into a moth., 4. Students may say that they read about metamorphosis. Some students may have had the same experience as described in the story. (You may have worked with pages 81 and 82 in this book.), 5. Situation1: Responses should suggest isolating the variable believed to be the cause of the caterpillar growth and controlling all other variables. Situations 2 and 3: One "test" would be to attempt to have someone "on watch" at all times to actually see the caterpillar harden into the sac and the moth emerge from the sac. Another test could involve running a video camera at the approximate time the caterpillar is expected to change and to become a moth., 6. Responses will vary. Check to see the bases for students' inferences are valid.

P. 102 1. The algae will thrive on the phosphates. As the algae die, bacteria will use up the oxygen to break down the dead algae. If the bacteria use up too much oxygen, the perch will all suffocate and die., 2. From the diagram, students can see that algae and perch are in the same ecosystem. Students have read that phosphates cause algae to grow faster., 3. Students can do an experiment by growing algae in one tank with phosphates mixed in and in one without phosphates., 4. There will be more perch and frogs for a while because fewer will be eaten. Because there will be more frogs and perch, more insects will be eaten, so there will be fewer of them., 5. From the diagram, students know that herons eat perch and frogs and that frogs and perch eat insects., 6. Responses will vary. Be sure students include the idea that they used knowledge available to them and applied it in a logical way to each situation.

Pp. 103-104

ACROSS: 2. insects, 3. starfish, 6. environment, 8. Gills, 9. lobster, 13. skeleton, 14. mammals, 15. classes, 16. birds, 17. backbone, DOWN: 1. nerve cord, 4. amphibian, 5. Fish, 6. earthworm, 7. mollusks, 10. reptile, 11. reproduce, 12. Scales

P. 112 1. They have the same types of foods grouped together., 2. The food pyramid shows five basic food groups; the old food groups had only four different groups. The fruit and vegetable group has been divided into two groups, the fruit group and the vegetable group. The old food groups do not have a place for fats, oil, and sweets. The number of food servings in each group is different., 3. Scientists and nutrition experts have made new discoveries about food and nutrition and about how much we should eat from each group., 4. If you start at the bottom of the pyramid, it's easier to tell what foods you should eat the most of. It's easier to plan a balanced diet by using the food pyramid., 5. Choices should be made after considering what was or will be eaten for the other meals of the day. Remind students that they can combine items from two or more groups, such as a bagel with milk.

P. 114 ACROSS: 2. Carbohydrates, 3. tomato, 6. vitamins, 7. fish, 8. spinach, DOWN: 1. protein, 4. milk, 5. oil, 7. fats

P. 115 Carbohydrates: cereal, bread, potato; Fats: nuts (can also be protein); Vitamin C: orange, strawberries; Protein: egg, steak, nuts; Vitamin A: carrots, peach; Calcium: cheese, milk

P. 117 Circle milk, toothbrush, girl brushing, fruit, girl at dentist; X on gum

P. 119 X on girl sneezing on boy

P. 122 1. two months, 2. d.p.t., or diphtheria, whooping cough, tetanus, 3. German measles, measles, mumps, Student responses with regard to their own vaccination records will vary.

P. 124 1. 90 cm (35 in.), 2. 20 cm (9 in.), 3. about 95 cm (about 38 in.), 4. about 5 cm (about 2 in.), 5. The student grew more between the first and second years of age., 6. 120 cm (47 in.), 7. The student grew 10 cm (4 in.) between 4 and 5 years of age., 8. The student grew the most during the first and second years of age., 9. The student grew 90 cm (35 in.) to 135 cm (53 in.).

Pp. 125-126

1. Astronauts need to exercise because when they are in space, their bodies do not work against the full force of gravity and their bones and muscles get weak., 2. The treadmill makes their arm and leg muscles work harder., 3. While pushing against the wall, you do not move forward. But you do exercise your legs and heart by working against resistance., 4. Carlos, 5. Carlos, 6. Carlos'

P. 128 You need more oxygen when you exercise.

P. 130 1. 200, 2. about 500, 3. on March 4, 4. The number of pollen grains in the air decreases because the rain washes some of the pollen to the ground., 5. Pollen grains could be carried from nearby woods and fields by the wind., 6. There would be fewer pollen grains in the air because deserts have less vegetation than other types of land.

P. 131 Students may draw cars, trucks, factories, etc. They may suggest walking or riding their bikes when they can instead of asking for a ride. They may mention carpooling.

P. 132 1. sunscreen, 2. hat, 3. hazy, 4. damage, 5. aspirin

P. 134 A.. Horsefly, B. Honey Bee, C. Mosquito, D. Wasp

Pp. 135-136

1. They hatch from eggs laid in the water., 2. It looks like a tiny caterpillar., 3. Wrigglers eat tiny plants and animals found in the water., 4. The mosquito picks up disease-causing organisms from an infected person's (or animal's) blood and passes them to the next person (or animal) it bites., 5. The mouth parts are sharp and can cut the skin., 6. Other wildlife could be harmed., 7. the spreading of natural enemies or diseases that kill mosquitoes

Pp. 137-138

A. Poison Ivy, B. Poison Oak, C. Poison Sumac

P. 139 1. outdoors, 2. Poison ivy, 3. Calamine, 4. poison sumac, 5. Poison oak, 6. plants, 7. sap, 8. rash, 9. berries

P. 140 Answers will vary.